SPATIAL TRANSPORTATION MODELING

CHRISTIAN WERNER
School of Social Sciences
University of California, Irvine

SAGE PUBLICATIONS
The Publishers of Professional Social Science
Beverly Hills London New Delhi

For information address:

SAGE Publications, Inc.
275 South Beverly Drive
Beverly Hills, California 90212

SAGE Publications India Pvt. Ltd.
M-32 Market
Greater Kailash I
New Delhi 110 048 India

SAGE Publications Ltd
28 Banner Street
London EC1Y 8QE
England

Printed in the United States of America

International Standard Book Number 0-8039-2738-X
0-8039-2503-4 (pbk.)

Library of Congress Catalog Card No. L.C. 85-062289

FIRST PRINTING

CONTENTS

INTRODUCTION TO THE SCIENTIFIC GEOGRAPHY SERIES

Scientific geography is one of the great traditions of contemporary geography. The scientific approach in geography, as elsewhere, involves the precise definition of variables and theoretical relationships that can be shown to be logically consistent. The theories are judged on the clarity of specification of their hypotheses and on their ability to be verified through statistical empirical analysis.

The study of scientific geography provides as much enjoyment and intellectual stimulation as does any subject in the university curriculum. Furthermore, scientific geography is also concerned with the demonstrated usefulness of the topic toward explanation, prediction, and prescription.

Although the empirical tradition in geography is centuries old, scientific geography could not mature until society came to appreciate the potential of the discipline and until computational methodology became commonplace. Today, there is widespread acceptance of computers, and people have become interested in space exploration, satellite technology, and general technological approaches to problems on our planet. With these prerequisites fulfilled, the infrastructure needed for the development of scientific geography is in place.

Scientific geography has demonstrated its capabilities in providing tools for analyzing and understanding geographic processes in both human and physical realms. It has also proven to be of interest to our sister disciplines, and is becoming increasingly recognized for its value to professionals in business and government.

The Scientific Geography Series will present the contributions of scientific geography in a unique manner. Each topic will be explained in a small book, or module. The introductory books are designed to reduce the barriers of learning; successive books at a more advanced level will follow the introductory modules to prepare the reader for contemporary developments in the field. The Scientific Geography Series begins with several important topics in human geography, followed by studies in other branches of scientific geography. The modules are intended to be used as classroom texts and as reference books for researchers and professionals. Wherever possible, the series will emphasize practical utility and include real-world examples.

We are proud of the contributions of geography, and are proud in particular of the heritage of scientific geography. All branches of geography should have the opportunity to learn from one another; in the past, however, access to the contributions and the literature of scientific geography has been very limited. I believe that those who have contributed significant research to topics in the field are best able to bring its contributions into focus. Thus, I would like to express my appreciation to the authors for their dedication in lending both their time and expertise, knowing that the benefits will by and large accrue not to themselves but to the discipline as a whole.

—*Grant Ian Thrall*
Series Editor

SERIES EDITOR'S INTRODUCTION

Transportation modeling is both one of the valuable job skills offered by scientific geography and a topic that can serve to develop analytic intuition. This book is designed for the student receiving a first exposure to the transportation problem as well as an introduction to the formal modeling of geographic phenomena.

Professor Christian Werner has restricted the presentation of the transportation models to those that could be expressed using only mathematics normally expected of first-year university students. The text is organized in order of the sequence of steps generally practiced in urban transportation planning rather than by methodology: estimation of transportation flows, network loading, optimal transportation decisions through linear programming, and design and operation of networks.

Modeling approaches that according to past experience seem to be most readily accessible to beginning students are the trip distribution models based on growth factors (without iterations), and, as an introduction to linear programming, the graphical solution and stepping-stone method. Slightly more demanding are the shortest route algorithm, Goldman's problem, and the Law of Refraction in transportation planning. Intermediate levels of difficulty are presented by the intervening opportunity model, the simplex technique, and the second example of optimal route design. Most of the remaining models and methods are more tedious than they are difficult: Examples are the abstract mode model or the trip distribution iterations.

Transportation modeling is a good and particularly useful example of the sharing of paradigms and methodologies between scientific geography and other sciences. Professor Christian Werner's geographical approach should be of particular interest to students and followers of the literature not only in human geography but also in operations research, transportation engineering, urban and regional economics, regional science, city and regional planning, and management science.

—*Grant Ian Thrall*
Series Editor

SPATIAL TRANSPORTATION MODELING
An Introduction

CHRISTIAN WERNER
School of Social Sciences
University of California, Irvine

1. INTRODUCTION

Transportation is a rather conspicuous part of human activities: There is the hardware—trucks and airplanes, bicycles and bicyle paths, railway stations and canals and parking lots and large freeways and small driveways, and so on. The hardware is the means by which we move people and freight from place to place—that is, from where they currently are to where, by somebody's decision, they are supposed to be. This last point provides us with a convenient opening to the phenomenon of transportation. At its beginning we always find a human decision, and the act of transportation is simply the implementation of that decision.

Let us itemize the content of such a decision:

- What should be moved? Usually people or cargo.
- From where and to where should it be moved? The origin and the destination of the movement.
- What means of transport should be used? The choice of one or a sequence of transportation modes.
- What route should the transportation movement take? The path connecting the points of origin and destination.

Thus, children being sent to a local school is a transportation example quite different from the shipment of grain from the Midwest to the Soviet Union, but the formal categories listed above apply in both cases.

Let us now approach the matter of transportation from the perspective of human settlement. Wherever we find human habitation, we find people engaged in activities such as working, socializing, eating, sleeping and relaxing. Not all of these activities can be conveniently done at one location; different activities require different settings, tools, resources. Some areas are reserved for agricultural or industrial production; other areas, such as parks, serve recreation, while residential areas provide housing. Human activities must then be distributed over space, occupying different sites.

It is this spatial distribution of human activities and the differentiation in land use that guarantee the need for transport, either of people to places of particular activities, or of material goods needed in the pursuit of these activities. Examples include the daily journey to work from the residence to the place of employment, the shopping trip to the neighborhood store or shopping plaza, trips to business conventions or sporting events, trips to visit relatives or attend parties, and the large-scale movements of agricultural products, oil, and other raw materials to places of processing or consumption.

Differentiation of land use is only in part caused by the need of various human activities, such as shopping or raising cattle, for their own and separate space. People, for their material livelihood, depend on available material resources. However, many essential resources do not exist where they are needed. For example, the areas of fertile soils, coal deposits, fresh water, or sufficient precipitation are determined by nature. The spatial distribution of these resources dictates where many economic activities take place.

What has always emerged as a result of human settlement is a spatial pattern of diversified land use brought about by the existing distribution of natural resources, by the variety of human needs and wishes, and by the specialization of economic activity. Changing patterns of land use, in turn, are accompanied by the development of transportation systems providing movement of people and goods between different places in a continuous and repetitive pattern.

At least since the beginning of this century, transportation systems development has increasingly been guided by the principles of science. The following section will introduce modeling as a scientific method of analysis and prediction.

Models: Tools in the Study of Real-World Phenomena

Let us now reflect on the style of our exploration of transportation issues. The vocabulary we have used so far consists almost exclusively of concepts and categories rather than individual objects and their names, reflecting our attention to sets of things or activities rather than to single objects and events. We group individual things into sets because we recognize commonalities in the vast variety of individual phenomena and their interdependencies. Thus, we refer to transportation flow rather than student X driving to campus Y, and we refer to land use generating transportation flow rather than campus X being the destination of students and faculty. If all things were unique, there would be no science, because science consists of statements about sets of things and relations that have something in common, at least in our perception and the way we interpret them.

Most things and the surroundings in which they are embedded are too complex for our comprehension: We do not know why they are the way they are; that is, we do not understand them. We do, however, notice similarities, regularities, and repetitions in our observations, and we formulate them as statements of experience. For example: Things that exhibit certain properties may in turn also possess certain other properties; or, if certain circumstances apply, then certain events will typically take place. As an illustration, a traveler confronted with several routes to a destination will usually decide on one of the quickest routes. Another well-known example: The amount of transportation flow between population centers tends to increase with the size of the respective population figures and decrease with increasing distance between the centers; this relation is known as the "gravity" behavior of transportation flows.

These statements are generalizations obtained by induction, that is to say, by identifying regularities in a set of empirical data. Alternatively, we might proceed in the logically reverse direction of "deduction" by inventing general statements that make claims about how things work and how they are interrelated. Combining these statements according to the rules of logic will produce new statements. In either case we end up with statements resulting from a cognitive process applied to reality. This is, of course, the main business of scientists: observation, recording, comparison, inference, speculation, and, eventually, testing. Together, these steps are known as the modeling approach, and transportation modeling is one example of this scientific enterprise. This approach will briefly be outlined in the following section.

The real world is too complex for our immediate and direct understanding. Scientists therefore create imaginary worlds, or "models," that may have, or are hoped to have, some similarity with a part of reality. Scientists always portray a selected segment of the real world in the form of some rather simple scenario described by a set of assumptions, some of which may appear to be quite implausible in a real-world context. However, these scenarios are designed in such a way that the scientist is able to discover implications and consequences that result from the combination of these assumptions.

Whatever questions can be answered, and whatever valid conclusions can be drawn within the setting of such a model, they may lose their validity when applied to the real world. This outcome is not at all unexpected, because models always represent reality in a crude, incomplete, and mutilated fashion. Nevertheless, the conclusions derived within a model might reproduce or predict real-world data, and that is one way in which a model can be "successful." Fortunately, many events that occur in the real world seem to be largely the result of a limited number of major forces. If we succeed in accurately describing these forces and their relationship to the event in question as assumptions in our model, we are able to forceast the event.

The model provides us with a simulation of some segment of the real world, suggesting how it might work and how it might behave if certain conditions change. Many of the well-established modeling approaches have become essential tools in transportation planning, and the greater portion of this text will be devoted to the description of a representative sample.

Transportation Planning

People have many common needs and desires, and since they are social animals, they engage in joint efforts to produce solutions that meet the needs of many. Most activities and their associated forms of land use—teaching in colleges, religious services in churches, living in residential areas, working in factories—are joint endeavors, and so is the provision of transportation that links the areas of specialized land use, allowing each of us to participate in a host of different activities.

In a highly organized and diversified society such as ours, adequate transportation requires enormous resources (on the order of 20 percent of the gross national product) and sophisticated organization and management. Such expenditures call for intelligent planning, which in turn requires that we understand the principles that govern transportation phenomena, their causes, their dynamics, their distribution, and their impact in space and time.

Nowhere is the demand for transportation and the need for its provision larger than in urban areas; therefore, it is urban transportation planning that we will use as a framework to organize our presentation of spatial transportation modeling.

Planning means to choose a set of goals and to identify a set of actions that will lead to the realization of those goals. It includes taking inventory of the current situation, because it is the present conditions that need to be acted upon to achieve that desired future situation described by our goals. In particular, planning includes the description of the

TABLE 1.1 **Transportation Expenditures as Percentage of U.S. Gross National Product**

Year	*1958*	*1961*	*1964*	*1967*	*1970*	*1973*
Passenger transportation	10.3	10.7	10.5	10.2	10.7	10.8
Freight transportation	9.8	9.2	9.5	9.0	9.0	10.1
Total transportation	20.1	19.9	20.0	19.2	19.7	20.9

SOURCE: U.S. Department of Transportation 1977.

relationships among the components of that situation. Only by understanding the web of interrelationships governing the components can we intelligently manipulate the current situation to reach the desired situation.

Planning always refers to some future scenario. In the case of urban planning, the goals typically describe an urban area as it should be and as it should function in ten or twenty years. Like the provision of energy and water, of health facilities and police protection and jobs and healthy air and urban parks, transportation is one of the prominent objectives that make up the broad goals of the plan. Specifically, the goal of transportation planning calls for a system of facilities that will provide individuals and organizations with "reasonably" inexpensive, safe, and quick transportation between the desired points of origin and destination without endangering other objectives of the plan.

We can readily agree that transportation planning has noble intentions, but how are those intentions to be implemented? To begin with, we do not even know how much demand for transportation there will be, say, ten years from now, what the spatial distribution of land use and human activities will be, and what transportation modes will be required. Even if we had this information, by what criteria do we decide that the current transportation system is inadequate to handle that future demand, or what constitutes a cost-effective change of the system so that it will "adequately" accommodate the future demand?

Clearly, we have a long way to go from the description of particular objectives for future transporation to a step-by-step implementation of regulatory and investment decisions, knowing at each step where we are and what will happen as a result of its implementation, and thereby tightly controlling the gradual shift from the present to that future state of affairs called for in our plan.

Transportation is a critical service function facilitating the thousands of events and processes that together make up the daily life of an urbanized area. Changing the existing transportation system means changing the quality of access to different locations and eventually changing human welfare. Some locations may become more desirable as a result of a change in the transportation system. In turn, this will lead to higher rents, to the outmigration of those who cannot afford to pay them, to changes in land use from, say, pasture to light industry or from low-income residential to office park. As the transportation system is changed to accommodate existing or projected transportation demand by extending the freeways or the public transit network, we bring about changes in accessibility, in land values, in land use, and therefore again a change in the amount and spatial distribution of demand for transportation.

In computer language the feedback process described here is called a "loop." The output of the loop, namely the adjustment of the transportation system, in turn changes the input to the loop for the next round, namely the demand for transportation. Usually, several repetitions of the sequence—transportation demand, systems adjustment, new transportation demand, new adjustment—will quickly converge to a final (stable) transportation demand pattern. Transportation planning is therefore not merely reactive or geared toward the accommodation of expected transportation demand but, through its impact on the other sectors of urban activities, can be used to bring about desired land use changes.

In this text we will consider the goals of urban planning and those of its transportation component as given; it is up to the urban society and its political representatives to decide what is the most desirable use of income and tax dollars. How a society wishes to live and shape its future is a political, not a scientific, matter. Rather, this text addresses the question of what needs to be done to achieve stated objectives or to determine whether stated objectives are achievable. The methodology that has been developed in the scientific pursuit of the spatial aspects of this question is known as spatial transportation modeling: the study of the spatial distribution of transportation phenomena with the help of models.

We begin with the dissection of the transportation planning process into manageable segments. The transportation system as planned for some future time has to accommodate the demand for transportation at that time; the demand, in turn, is the direct result of the future pattern of land use. At this point we assume that economic and population planners have already projected the future land use pattern with the help of their own forecasting models (not to mention development decisions already made and awaiting implementation). It is the role of the trans-

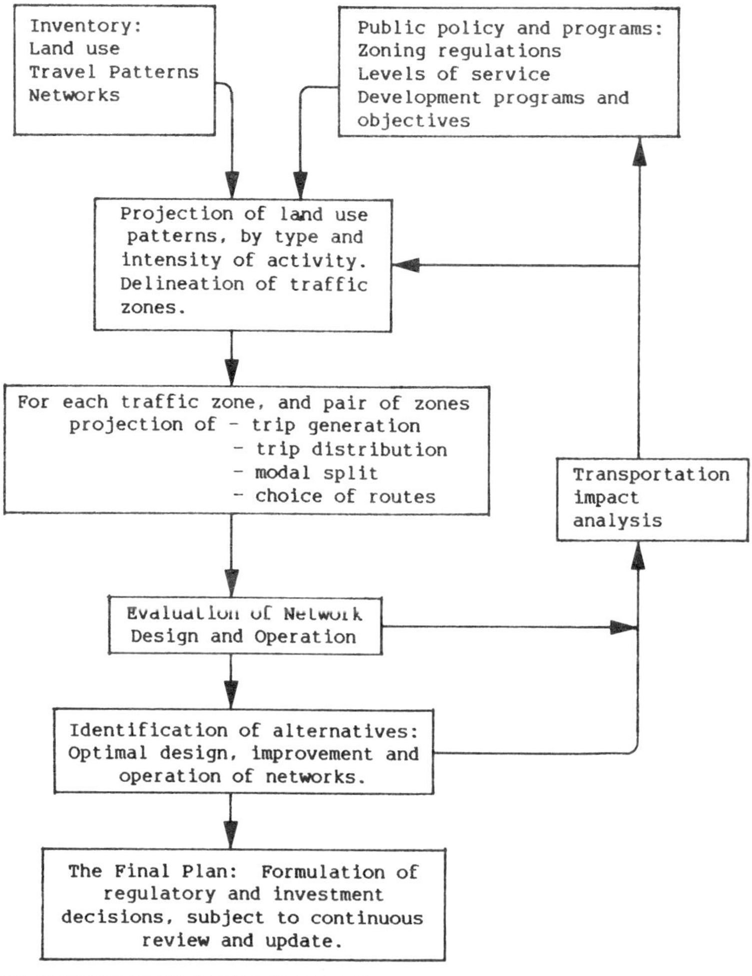

Figure 1.1 Simplified Flow Diagram of the Transportation Planning Process

portation planner to design a transportation system that will provide adequate service. A flow chart summarizing the transportation planning process is provided in Figure 1.1.

First, the transportation planner subdivides the planning area into a set of subareas called traffic zones. Often the boundaries of these zones are dictated by the spatial breakdown of the statistical data bases available, but they might also be delineated in preparation for a transportation plan. Using models estimating the quantitative relationships between type of land use and traffic generation ("trip generation models"), the planner then determines for each traffic zone the transportation demand it will generate, either as movement to other areas (that is, outflow) or as movement attracted from other areas (that is, inflow).

Second, having estimated the demand generated by each land use area, the transportation planner needs to know what the destinations of the outflows and the origins of the inflows are. There is a set of models, referred to as "trip distribution models," that use a variety of information about the urban area and its subareas to establish the distribution in terms of origins and destinations for the transportation demands forecasted by the trip generation models.

Third, what transportation modes will those anticipated flows use: private car, train, bus, or truck? Again, transportation scientists have developed several models, referred to as "modal split" or "modal choice" models, that estimate the distribution of the transportation flows over the various transportation modes.

Fourth, what route will the transportation flows take? In an urban area, ground transportation accounts for virtually all transportation flows, and these flows always take place along fixed routes, which together form transportation networks. To predict the route a transportation flow will take through a certain network, transportation scientists have developed several "network loading" procedures, also known as "trip assignment" models. These models are usually based on the assumption that network users will try to minimize the time it takes them to get to their destination by choosing the quickest route, taking into account delays resulting from congestion and intersections.

Loading the pattern of future transportation demands onto the current transportation network permits the transportation planner to simulate future network performance. The simulation will establish whether the transportation system is adequate according to preestablished performance criteria—for example, whether the average speed of the transportation flows is at least thirty miles per hour. In fact, many of the existing urban networks do not currently meet minimum performance standards, and their performance would be worse for the flow volumes projected for future years.

Fifth, how should the existing network be improved so as to meet the future anticipated demand? Once again, transportation science provides a series of models evaluating alternative network designs; other models generate and evaluate different strategies of network operation and use: routes and schedules of public transit, control of network access, traffic regulations, traffic control in intersections, user charges, and the like. These models are necessarily normative: Instead of predicting what will happen, they tell the planner either what changes in the network should be implemented to minimize the associated costs while meeting pre-established performance standards, or how to achieve maximum performance for a given budget.

Conclusion

Changing the transportation system will change the quality of access to various locations, inducing changes in land values and land use, and reducing transportation cost, which will mean savings for the urban economy. These savings might reappear as additional economic demand inducing added economic growth, which generates additional transportation flows as well as a host of other consequences. These consequences have traditionally been underestimated or altogether ignored by planners, at least in part because they are difficult to trace, measure, and separate from other developments unrelated to transportation system changes. Recent studies of transportation impacts have clearly established the significance of these chain reactions as they reverberate through the complex urban system. Estimation methods, however, are still in their infancy, and they will be only briefly discussed in this text.

Finally, there are several feedback processes at work between the dynamics of a transportation system and the environment in which it is embedded; some of these have been well documented. One example is the chain that leads from traffic congestion to network improvement to improved accessibility to increased economic activity to new traffic flows and new congestion. As long as those processes are not successfully modeled, they will continue to be one of the biggest sources of error in the field of medium- and long-run transportation planning.

The organization of this text will closely follow the sequence of steps of the transportation planning process reviewed above.

Chapter 2 presents a series of models estimating the demand for transportation generated by urban areas, and the disaggregation of this demand by destination and mode of transportation.

Chapter 3 introduces basic concepts from graph and network theory and presents a dynamic programming algorithm that determines the maximum flow or, alternatively, the shortest path through a network.

Chapter 4 discusses the principal elements of elementary Linear Programming by means of a graphical solution procedure. It reviews several applications of Linear Programming to particular transportation problems and concludes with a presentation of the simplex method for nondegenerate cases.

Chapter 5 explores the economic rationale underlying the form and function of transportation networks and presents examples of optimal route design and link addition.

Chapter 6 summarizes some of the methods designed to estimate the social and economic impact of transportation system changes on the environment in which it is embedded. It closes with an example of an econometric model designed to predict the impact of changing transportation cost on the performance of the national economy.

2. ESTIMATION OF TRANSPORTATION FLOWS

In this chapter we introduce models that estimate the generation and distribution of transportation flows, and their composition by mode of transport.

Trip Generation Models

Viewing city traffic from the air provides a convincing demonstration of the immense complexity of urban travel patterns, consisting of innumerable individual movements merging with and then leaving traffic flows, only to join other flows until they eventually terminate at some point in the urban landscape.

The transportation planning process presented earlier gives us an orderly and systematic framework for the detailed description, estimation, and prediction of urban traffic flow. The first step addresses the question of trip generation: What is it that causes trip making? What environmental circumstances lead to the production or attraction of traffic? If modeling is defined as a simulation of the cause-effect chains of real-world events, then we are still largely without models, at least models that have survived the test of application in a real-world context. Instead, planners take advantage of observed covariations between socioeconomic characteristics of an area and the number of trips generated in that area.

As residential areas generate between 80 and 90 percent of the trips in urban areas, we will review models estimating the numbers of trips originating or terminating in residential areas. The basic trip-generating unit in such areas is the individual household. The number of trips it generates has been shown to be statistically dependent on family size, on the number of cars owned, and on income, among other characteristics. Gathering and processing data from sample households permits the establishment of an equation in which number of daily trips is expressed

TABLE 2.1 Origins and Destinations of All Internal Trips, by Trip Purpose, in the Chicago Area, in Percentage of Total Trip Number

Origin \ Destination	*Home*	*Work*	*Shop*	*School*	*Social/ Recreat.*	*Personal Business*	
Home	–	17.8	5.3	2.0	9.9	8.1	43.1
Work	17.0	3.1	0.9	–	0.3	0.4	21.8
Shop	5.8	0.6	0.7	0.1	0.5	0.4	8.1
School	1.7	0.1	–	–	0.1	–	1.9
Social/Recreat.	11.3	–	0.6	–	1.7	0.4	14.2
Personal Business	7.4	0.3	0.6	–	1.0	1.6	10.9
	43.2	21.9	8.2	2.1	13.6	11.0	100.0

SOURCE: Chicago Area Transportation Study 1959.

as a function of various household characteristics (McCarthy 1969; Stopher and Meyburg 1979).

As an example, consider the number of trips per household as a function of family size. For each of the households in the sample, a figure of the daily number of trips and of the number of household members is obtained. To illustrate the derivation of the functional relationship between the two variables, we plot the data on graph paper organized by a system of coordinates. Trip numbers are measured along the y-axis, and size of household along the x-axis. Each sample household will appear as a point in the plot. Together, these points tend to form a fairly "straight" pattern, that is, a pattern into which we can fit a straight line that constitutes a reasonable representation of the data. This line is called a regression line, and the algebraic expression $y = a + bx$ is called a regression equation.

The line is usually fitted by what is called the least squares technique. The parameters a and b are calculated under the following condition: Let x', y' be the coordinates of a sample point, and let y be the value of the regression line for the same x'. Then $|y - y'|$ measures the deviation of the sample point from the regression line. The particular line for which the sum of the squares of these deviations summed over all sample points is minimized, is known as the least squares regression line (Yule and Kendall 1965). The mathematical derivation of this method is based on calculus and will not be presented here.

Multiple regression equations are based on the same principle: They express some variable as a linear function of two or several other variables. Again, the equation parameters are estimated with the help of actual survey data and the least squares technique.

A major purpose of regression equations is to estimate variables difficult to measure, with the help of other variables that can be more easily obtained or are already available. It should be clear that the qualilty of the estimates tends to be better the closer the sample data scatter around the regression line.

The following multiple regression equation is based on data from the Washington, D.C., metropolitan area; the equation estimates the daily number of trips T per household for a given residental area (Mertz and Hamner 1957):

$$T = 4.33 + 3.89\ x_1 - 0.005\ x_2 - 0.128\ x_3 - 0.012\ x_4$$

where x_1 is the number of cars per household, x_2 is the number of households per acre, x_3 is the distance from the central business district in miles, and x_4 is the income per household in thousands of dollars.

Suppose that we have generated estimates of the number of household trips as a function of household size, income, number of cars in the household, and other characteristics. We multiply each of the estimated trip numbers by the number of households that have the corresponding characteristics. The sum of all of these products constitutes the number of trips generated in the area under consideration. Alternatively, one can express the total number of trips generated in a particular traffic zone as a function of the total population, the number of registered cars, and total income in that zone. The equation used to estimate the number of trips generated in zones with nonresidential land use would be based on other variables that correlate highly with the number of trips generated, such as employment figures (in industrial areas) or number of retail establishments (in shopping areas).

The application of regression analysis carries with it a host of assumptions and prerequisites, such as the independence of the variables and linearity of relationships between dependent and independent variables. Moreover, the variables have to be distributed approximately normally, and the data sample has to meet certain conditions of random sampling. Many of these requirements cannot, or at least cannot easily, be met, and this may in part account for the sometimes disappointing performance of this particular approach.

Needless to say, poor estimates of current trip generation figures become worse when the same regressions are run for some future date. To begin with, estimates of variables such as households or automobile ownership are prone to significant error; furthermore, applying regression equations calibrated with current data to some future point in time assumes stability of that relationship over time. In other words, it assumes that conditions which prevailed in the past and resulted in the particular set of data used to estimate the equation parameters will prevail in the future. This is the familiar assumption of parameter stability. In contrast, history teaches us that the ongoing changes in the way individuals and society set their priorities and attend to their needs make it unlikely that parameters remain stable over any but the shortest time intervals.

Another approach to forecasting trip generation of residential areas is a method known as category analysis (Wotton and Pick 1967). The households of a survey sample are grouped by selected categories such as family size, income, and number of cars. Each of these categories is broken down into classes such as ownership of no cars, one car, or two or more cars. Thus, one group may be defined as all households in the sample with a family size of three or four, a household income of between $15,000 and $20,000, and with two cars. Using the data from the same survey, the average number of daily trips per household is determined for each group.

Then, for some target date in the future, projections are made as to the number of households in a particular traffic zone and their breakdown into the groups described above (these projections are prepared by other segments of the urban planning process). Multiplying the number of households in each group by the number of daily trips characteristic for that group and summing over all groups will give the planner an estimate of the number of trips generated in that zone at the time of the target date.

The susceptibility of this method to various types of errors is obvious: Once again, there are the assumptions that the relationship between some socioeconomic properties and trip-making behavior will remain stable over time, and that the projections of the number and types of households in a given area at some future time can be made with adequate accuracy.

Trip Distribution Models

The trip generation models presented above only estimate the number of trips having one of their trip ends in a particular traffic zone. It is the purpose of trip distribution models to identify the other trip ends, that is, the destinations of the trips originating in a given traffic zone or the origins of the trips attracted to that traffic zone. The following sections will review several distribution models.

The need for methods estimating the distribution of trips has grown with the increase of both the population and its mobility. In the absence of planning, the demand for transportation was often met simply by reacting to severe congestion, to unusually high accident figures, or to economic stagnation apparently resulting from lack of access. Transportation improvements through alleviation of individual emergencies are solutions after the fact and have all the attributes of patchwork: They frequently create bottlenecks elsewhere, or they are out of date by the time they have been implemented. What the transportation planner needs to know is the transportation demand pattern in the years to come so as to be able to anticipate inadequacies of the transportation system and take corrective measures ahead of time.

A particular flow pattern might result from many individual and independent decisions, as in the case of shopping trips in an urban area. Alternatively, it may be the outcome of an organized set of decisions made by a single authority. An example of the latter would be an oil company deciding on the schedule of gasoline shipments from its various refineries to its gas stations.

While it is exceedingly difficult to model many independent decisions with uncertain objectives, it is often possible to make accurate predictions for a single decision when the objective of the decision is known. In the case of an oil company, such an objective may be the minimization of transport cost or time, or the maximization of safety or profit. The modeling of single transportation decisions with known objectives is treated in the fourth chapter of this text; in this chapter we review a selection of models that estimate transportation flow patterns representing large numbers of individual trip-making decisions.

Growth Factor Models

The earliest and simplest approaches to the estimation of the number of future trips between different zones are the so-called growth factor models (Martin et al. 1961). They are based on the assumption that interzonal flows will grow either at the same rate as the total trip number generated in the overall area under consideration (usually a larger municipality or a metropolitan area) or at a rate that depends on the growth of traffic generated in individual traffic zones. The simplest version of these models assumes the future flow volume T_{ij} between traffic zones i and j to grow at the same rate as the traffic for the entire area:

$$T_{ij} = t_{ij} \cdot G \qquad [1]$$

where T_{ij} is the trip volume between i and j projected for some future time, t_{ij} is the current trip volume, and G is the growth factor, defined as the ratio of the projected and the current number of trips in the entire area: $G = T/t$ where T has been estimated with the help of a trip generation model and t is based either on survey data or on the application of a trip generation model to current land use data.

Notice that equation 1 projects all future interzonal flows by using the same constant multiplier G, ignoring the fact that different zones typically experience differential growth rates of population and economic activity, which in turn leads to different growth rates for the interzonal flows. Unlike other models, however, this model is internally consistent inasmuch as its output figures T_{ij}, for all pairs of traffic zones i, j, conform to its input figures t_{ij} and G. That becomes clear if we sum the projected flows for the entire area while keeping in mind that each flow is counted twice in the double summation, as T_{ij} and T_{ji}:

$$\frac{1}{2}\sum_i\sum_j T_{ij} = \frac{1}{2}\sum_i\sum_j t_{ij}\frac{T}{t} = T\,\frac{\frac{1}{2}\sum_i\sum_j t_{ij}}{t} = T. \qquad [2]$$

The relation formulated in equation 2—that the sum of the individual flows is equal to the total flow volume—is known as the first conservation rule of traffic flows. Notice that the T_{ij}, on the left side of equation 2, is an estimate made by the model, while the figure T, on the right side, is part of the model input. As we will see in the next example, certain models will generate output figures that are in contradiction to the input on which they are based.

To get away from the assumption of a uniform growth rate G of trip numbers throughout the area and to make individual trip projections sensitive to the differential growth in different zones, the projected number of trips T_{ij} has been estimated in terms of the particular growth rates in the zones of trip origin and destination, i and j:

$$T_{ij} = t_{ij}\,\frac{G_i + G_j}{2} \qquad [3]$$

where $G_i = T_i/t_i$ and $G_j = T_j/t_j$ are the growth rates of trips generated in i and j; t_i, t_j and T_i, T_j are the current and projected trip volumes generated in zones i and j; and

$$t_i = \sum_j t_{ij}, \quad t_j = \sum_i t_{ij}.$$

Unfortunately, this model does not satisfy the second conservation rule: The sum of the projected number of trips leaving or terminating in i, ΣT_{ij}, usually differs from the (input) number T_i of trips generated in i. The following consideration will demonstrate this point: We select the zone i with the lowest growth rate:

$$G_i = \text{Min}\{G_k | k = 1, 2, \ldots\}. \quad [4]$$

In words: G_i is the minimum value of the growth rates of all traffic zones. The projected number of trips originating or terminating in i is

$$\sum_j T_{ij} = \frac{1}{2} \sum_j (G_i + G_j) t_{ij} > \frac{1}{2} \sum_j (G_i + G_i) t_{ij}$$
$$= G_i \sum_j t_{ij} = \frac{T_i}{t_i} \sum_j t_{ij} = T_i. \quad [5]$$

Consequently, $T_i < \sum_j T_{ij}$, when the two terms should, of course, be equal.

Another shortcoming of this model is its insensitivity to the growth that takes place in some zones other than i or j, despite the fact that such growth will usually have some impact on the number of trips between i and j. The Detroit model (Bevis 1956; Stopher and Meyburg 1975) tries to alleviate this problem by estimating the future flow between i and j, T_{ij}, as:

$$T_{ij} = t_{ij} \cdot \frac{G_i \cdot G_j}{G}. \quad [6]$$

Note that the estimate T_{ij} is now a function not only of the growth rates at i and j but also of the overall growth rate G in the entire urban area. If this rate is large relative to the growth rates at i and j, then relatively more trips originating or terminating in i or j will have their other trip end in zones other than i or j, thereby reducing the flow between i and j.

This model violates all conservation rules; the estimates it produces usually do not match up with the input figures of number of trips generated in the individual zones or in the entire area. An iterative procedure has therefore been devised which, through stepwise readjustment of the flow estimations and associated growth rates, will eventually balance the projected trip generation figures T_i, T_j and the total projected trip number T with the forecasted interzonal flows T_{ij}.

The procedure is quite straightforward: Apply the model to the current trip numbers t_{ij} and the projected trip generation figures T_i, T_j and

$$T = \frac{1}{2} \sum_i T_i = \frac{1}{2} \sum_j T_j.$$

This step, called the first iteration and identical to equation 6, will yield a set of estimates T_{ij}. Next, apply the model to these estimated figures to generate a set of refined estimates. This step constitutes the second iteration. The new estimates are further refined by additional iterations until the discrepancy between model input and output is judged to be sufficiently small.

As a simple example of the iterative procedure, consider an area subdivided into three transportation zones, 1, 2, 3. Let the current flows between them be $t_{12} = t_{21} = 3$, $t_{23} = t_{32} = 2$ and $t_{31} = t_{13} = 1$. Thus, the current trip generation figures are $t_1 = t_{12} + t_{13} = 4$, $t_2 = t_{23} + t_{21} = 5$, and $t_3 = t_{31} + t_{32} = 3$. The total number of current trips is $t = t_{12} + t_{23} + t_{31} = 6$. For some particular future time, the trip generation figures of the three zones have been projected to be $T_1 = 6$, $T_2 = 5$, and $T_3 = 9$; there is no growth predicted for zone 2 and a tripling for the volume of traffic generated in zone 3. The total projected trip number is therefore $(T_1 + T_2 + T_3)/2 = 10$, and the overall growth rate is therefore $G = T/t = 10/6$.

Applying the model once (which constitutes the first iteration) will yield the following estimates: $T_{12}(1) = 2.7$; $T_{23}(1) = 3.6$; $T_{13}(1) = 2.7$. That is, of the total number of future trips $T = 10$, only 9 have been allocated. However, the basic pattern of future flow projections is already apparent: The stronger growth in zone 3 leads to major increases of the flows between it and the other two zones, while the third flow, between zones 1 and 2, actually declines. After the fifth iteration, the figures are $T_{12}(5) = 1.42$, $T_{23}(5) = 3.675$, and $T_{31}(5) = 4.81$. The sum of the estimates is 9.98, that is, almost precisely the projected total trip number of 10 for which the model was supposed to determine the interzonal distribution.

Like its predecessors, this model has serious shortcomings: For example, at least in its present form, it cannot distinguish between flows from i to j and from j to i because its estimating equation is symmetrical in i and j. Also, the model cannot accommodate undeveloped traffic zones earmarked for future growth because there may not be any current flows to and from these areas that could be adjusted by means of growth factors.

Intervening Opportunity Models

Individual trip-making behavior depends not only on the attractiveness of the intended destination and the distance that has to be overcome to get there but also on the availability of opportunities elsewhere, opportunities that are competitive and can satisfy the purpose of the planned trip.

The concept of intervening opportunity is a familiar one. An example from geographic theory is Christaller's central place model (see King 1984, in this series). New merchants locate their businesses so that they can attract some of the customers of business establishments already in existence through greater proximity. However, the central place model is strictly deterministic, assuming as it does that customers will always choose the nearest opportunity, in contrast to observed human behavior. The concept of intervening opportunity is widely applied in decision making: Private enterprise and even governmental agencies study the quality and distribution of supply and demand and establish themselves in positions or under conditions of "intervening" opportunity, be it by offering easier access, cheaper prices, or better service, thereby diverting part of the existing consumer flows away from their former destinations.

Stouffer (1960) was apparently the first both to recognize the significance of the existence of intervening opportunities when individuals decide on a trip destination and to operationalize this concept. In estimating flows of job-seeking migrants, Stouffer was able to show that the number of job opportunities "between" the origin and destination of a flow had a negative impact on the flow volume, and that a more accurate flow estimation was possible if one used a gravity-type model, in which distance between origin and destination was replaced by the number of intervening opportunities.

At least implicitly, Stouffer's model addressed an aspect of human behavior that has made estimating and forecasting travel behavior painfully difficult: Human decision making does not seem to follow any simple, optimizing pattern. In particular, people do not always choose the closest supply place, the lowest bidder, or the fastest route, and it may not at all be simply for reasons of incomplete information about available choices, but for reasons the researcher cannot identify or is unable to measure. Moreover, opportunities and attractions are only that: They influence but do not determine human decision making—at least not in any way we know how to model, let alone understand. This dilemma has given rise to a proliferation of stochastic (probabilistic) models that allow for a variety of possible decisions with differing probabilities. The following model, first developed for the Chicago Area Transportation Study (1960), will illustrate the stochastic approach.

Consider an individual Z visiting a particular type of establishment, say a tire store of which there are many in Z's local area. We will assume that anyone will do for the purpose of Z, that Z is therefore indifferent and would choose anyone of them with the same probability p if their location did not matter. Location does matter, however, and we will therefore assume that Z considers the various alternatives sequentially according to their proximity to Z's location. Let P_k denote the probability that Z will decide on the kth closest establishment. Thus, Z chooses the closest establishment with probability $P_1 = p$; the probability that Z does *not* choose this store is therefore $1 - p$. The probability P_2 that Z chooses the next closest establishment is equal to the (constant) probability p of choosing it if location did not matter times the probability of not choosing the closest store, $1 - p$: $P_2 = (1 - p) \cdot p$. Likewise, the probability P_3 of selecting the third closest store is equal to the probability of not stopping at the first store, $1 - p$, times the probability of not stopping at the second store, again $1 - p$, times the (constant) probability p of choosing the third store: $P_3 = (1 - p)(1 - p) \cdot p = (1 - p)^2 \cdot p$. In general, the probability P_k of choosing the kth store is equal to

$$P_k = (1 - p)^{k-1} \cdot p. \qquad [7]$$

Unlike a deterministic normative model, this approach does not assume that a customer always decides on the closest opportunity; rather, the closest opportunity has only the highest probability of being chosen, and the other opportunities, further away, have decreasing probabilities of being selected: For $p = 1/3$, for example, it is

$$P_1 = \frac{1}{3} \text{ or } \frac{9}{27}; \quad P_2 = \frac{2}{9} \text{ or } \frac{6}{27}; \quad P_3 = \frac{4}{27};$$

and so on.

Since Z will eventually decide on a particular establishment and against all others, the choices available to Z are mutually exclusive: Choosing one eliminates the choice of any other. We recall from elementary probability theory: The probability that exactly one of n mutually exclusive events $x_1, x_2, \ldots x_n$ will take place is equal to the sum of the respective probabilities of these events:

$$P(x_1 \text{ or } x_2 \text{ or } \ldots x_n) = \sum_{k=1}^{n} P(x_k). \qquad [8]$$

This sum is equal to one if $x_1 \ldots x_n$ constitute all possible events. (For example, if we toss a die there is a total of $n = 6$ different events or outcomes; they are mutually exclusive; and the probability of the outcome being 1 or 2 or . . . or 6 is equal to $1/6 + 1/6 + \ldots + 1/6 = 1$, that is, certainty.)

Assume that Z is confronted with n mutually exclusive choices; the probability that Z will pick the first or the second or any other one, including the nth, must be equal to 1 because we have assumed that Z will eventually settle on one (and only one) of the various stores available. Thus, applying equations 7 and 8,

$$\sum_{k=1}^{n} (1-p)^{k-1} \cdot p = 1. \qquad [9]$$

Let S(n) denote the sum on the left side; this sum is a finite geometric series with constant multiplier $(1 - p)$ and is equal to

$$S(n) = \sum_{k=1}^{n} (1-p)^{k-1} = \frac{1-(1-p)^n}{p}. \qquad [10]$$

This becomes immediately apparent if we calculate $S(n) - (1 - p)S(n)$:

$$S(n) - (1-p)S(n) = \sum_{k=1}^{n} (1-p)^{k-1} - \sum_{k=1}^{n} (1-p)^{k}$$

or

$$p \cdot S(n) = \left[\sum_{k=1}^{n} (1-p)^{k-1} - \sum_{k=2}^{n+1} (1-p)^{k-1}\right]$$

$$= [(1-p)^0 - (1-p)^n].$$

Substituting equation 10 into equation 9 makes it clear that the expression $(1 - p)^n$ has to be equal to 0; that, however, is possible only if either p is equal to 1 or n approaches infinity. The first solution is incompatible with the purpose of our model because it implies $P_1 = p = 1$, meaning that Z chooses the closest store with certainty, ignoring all other alternatives. In this case the concept of intervening opportunity would not apply and the model would be of the conventional deterministic design. On the other hand, the assumption of an infinite number of stores from which to choose seems to render the whole approach untenable, but that is not the case. Remember that models are, at best, good approximations of reality, and to represent a large number of choices in reality by an infinite number of choices in theory is not at all a bad approximation, as the following example will show.

Assume that the number of stores, n, is equal to 4, and p is equal to 1/2. Then, according to equation 7, Z will choose the closest store with probability 1/2, the second closest store with probability 1/4, the third with probability 1/8, and the fourth with probability 1/16. If we add these probabilities, we get 15/16, a value already very close to 1. If the number n of stores from which to select is larger than 4, then the sum of the probabilities will be that much closer to 1 ($1 - 1/2^n$, to be precise). If the value of p is higher, say 2/3, then the first three stores would be chosen with probabilities 2/3, 2/9, and 2/27, or with a joint probability of 26/27, again a value very close to 1 and already on the basis of only three stores from which to choose.

Nevertheless, the left side of equation 9 will always be less than 1 because we may be dealing with, at most, a substantial number of establishments but not an infinite one. Thus, this model can be an approximation only on logical grounds—let alone its other shortcomings (for example, the assumption that each establishment is equally acceptable to the customer if distance is disregarded). Properly chosen, however, idealization and simplification are assets rather than drawbacks of a model, because they make it operational while maintaining sufficient similarity with the real world. We will therefore continue with the development of the model at hand.

We now introduce two extensions to the model as formulated in equation 7: We consider an entire urban area broken down into several individual zones, each zone being the origin of trips and each zone being also the location of a number of retail establishments. Let O(i) be the number of shopping trips originating in zone i, and let A(j) be the number of stores located in zone j. Based on our modeling approach developed so far, we will derive an estimate of the number of trips T(i, j) originating in i and terminating in j.

We order the different zones according to distance (or travel time) from zone i, with zone 1 being the one closest to i. We assume that an individual Z living in i will examine the available shopping opportunities sequentially and eventually choose one, and only one, store. Hence, Z will choose a store from the first j – 1 zones or a store in j or one from the remaining zones. Each of these three possibilities has a certain probability, and, because of equation 8, the three probabilities have to add up to one. A combined probability of one indicates certainty, and according to our earlier assumption, it is indeed certain that Z will choose one, and only one, of the three possibilities.

Let V(j) be the number of stores up to zone j:

$$V(j) = \sum_{k=1}^{j-1} A(k).$$

Z will pass up the stores of the first j – 1 zones with probability $(1 - p)^{V(j)}$ and will therefore choose one of them with probability $1 - (1 - p)^{V(j)}$; Z will choose a store in j with probability P(j) and will choose a store beyond zone j with a probability equal to the probability of not choosing one of the first V(j + 1) stores, namely $(1 - p)^{V(j+1)}$. Hence

$$[1 - (1 - p)^{V(j)}] + P(j) + (1 - p)^{V(j+1)} = 1 \qquad [11]$$

or

$$P(j) = (1 - p)^{V(j)} - (1 - p)^{V(j+1)}. \qquad [12]$$

Since O(i) is the total number of shopping trips originating in i, the number of those going to j is

$$T(i,j) = O(i)\,[(1 - p)^{V(j)} - (1 - p)^{V(j+1)}]. \qquad [13]$$

The following data requirements have to be met to apply this model in concrete situations: The number O(i) of shopping trips originating in zone i; the numbers A(k) of retail establishments for the various urban zones k = 1, 2, . . . , and the value of the constant probability p of a shopper choosing a store. Unfortunately, this value tends to vary with the purpose of the shopping trip: For ordinary groceries, almost any supermarket will do; therefore, p tends to be relatively high. When buying jewelry, on the other hand, shoppers are much more discriminating and willing to travel larger distances.

To improve the predictive power of the model, the shopping trips need to be disaggregated according to purpose, and the number of shopping trips from i to j would be the sum of the individual trip numbers predicted for different shopping purposes. Even then the model is still rather limited and needs to be broadened to account for multiple-purpose trips and trips that fail to reach their objective.

Gravity Models

Of the many planning models developed to estimate the number of trips between different areas, the family of gravity-type models represents the method most widely used. The simple gravity model assumes the number of trips between areas i, j to be proportional to the population figures in i and j and inversely proportional to the distance between i and j raised to some power. The analogy to Newton's law of gravity is obvious: This law states that the gravitational attraction between two bodies is proportional to their respective masses and inversely proportional to the square of the distance between them.

The early version of the gravity model has later been significantly improved. The population figures at i and j were replaced by the trip numbers produced in i and attracted to j; furthermore, for each pair i, j an empirically derived value measuring the degree of separation between i and j replaced the distance function, and a specific adjustment factor capturing social and economic influences on the trip pattern was included in the estimating equation. For a comprehensive coverage of gravity models and their applications, see Haynes and Fotheringham (1984, in this series).

Models of Modal Split

The application of trip distribution models to the trip generation data of an urban area will provide us with a so-called origin-destination matrix, that is, a table of trip data in which the number in the ith row and j^{th} column represents the number of trips from zone i to zone j (or, depending on the model, between i and j). This section will review a set of models that estimate, for each of these flows, the share of each transportation mode.

Urban transportation needs are being met by a number of different modes: Walking on sidewalks, mass transit by bus or train on networks of road and rail, transportation by bicycle or car, airborne transportation by commuter airlines or private aircraft. By far the most important modes are those of public transit and private automobile, important in that they account for the vast majority of trips in urban areas and create most of the urban transportation ills—congestion, pollution, accidents, costs. These problems can be contained, if not alleviated, by a variety of planning instruments: zoning regulations to control land use and thereby traffic generation, tolls and taxes to discourage use of private automobiles, subsidies for mass transit, or investments in highway network extension and improvement.

Whatever measures are taken, they have to take into account not only the anticipated volume and distribution of traffic but also its breakdown by mode. Early research efforts were somewhat successful in estimating the split of total number of trips in an urban area into those by public transit and those by private automobile; however, since these estimates refer to the entire urban area, they do not permit the design of transportation system improvements that would be sensitive to intraurban variations of modal split, which are considerable.

We will briefly review a few modeling efforts that estimate the share of each mode in the set of all trips between two traffic zones in an urban area. From a formal methodological standpoint, they do not differ from the models of trip generation we presented earlier, the two main methods, again, being regression and category analysis. Thus, the ratio of trips by transit versus those by automobile between two traffic zones would be expressed as a function of both socioeconomic and transportation-related parameters, the first set including employment, income, and population figures, and the second including figures on transportation costs and time for both transit and automobile between the two traffic zones. This regression approach faces the whole set of difficulties associated with sampling and statistical inference, which we encountered in the section on trip generation models.

The other method outlined in that section—category analysis—has also been used in the prediction of modal split, or modal choice, as it is often called. Empirical evidence from many studies indicates that the choice of individuals between public transit and their own car is heavily dependent on the ratio of the time it takes them by either transit or car to get to their destination, where time is measured as the duration of the entire trip—that is, from door to door. Plotting transit trips as the percentage of the total number of trips against the ratio of transit time versus automobile time produces a scatter diagram, and the curve fitted into that scatter is known as a diversion curve (Figure 2.1). Not surprisingly, the shape of a diversion curve depends on several other parameters as well: Public transit tends to attract a larger percentage of trips if automobile ownership in the zone of origin is low or if the monetary savings from using transit rather than the car are substantial. To capture the influence of these parameters, the trips between traffic zones are subdivided by trip purpose (say, work versus nonwork), by individual income, by the ratio of transit versus automobile costs of each

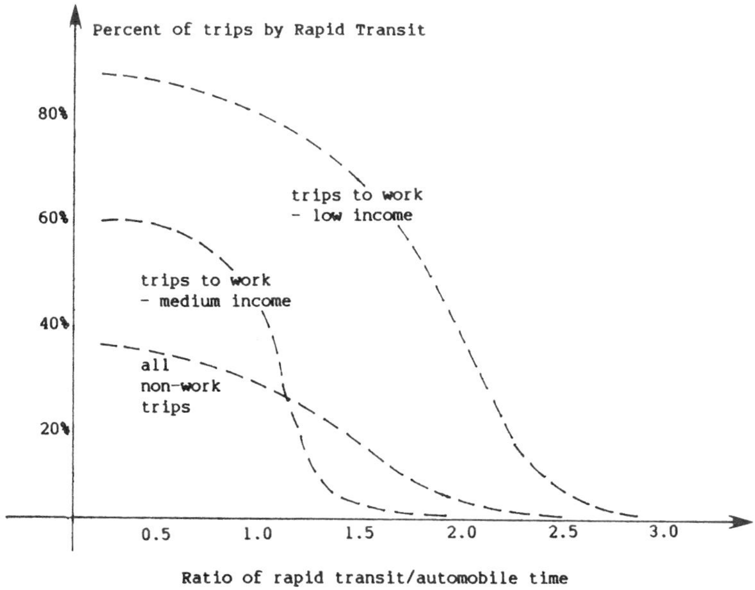

Figure 2.1 Examples of Diversion Curves Based on Travel Time Ratio

trip, and other trip characteristics. This multidimensional stratification process divides the trips into a number of classes.

An example of such a class might be the set of trips that are non-work-related, which cost twice as much by car as by transit and which are undertaken by individuals with incomes between $10,000 and $15,000. Based on survey data, a diversion curve is prepared for each class relating the percentage of transit trips to the trip time ratio of transit versus car.

The application of diversion curves to predict the percentage of transit trips between traffic zones for some future time proceeds in a fashion parallel to the use of category analysis in the estimation of trip generation (see the first section of this chapter). Other segments of the planning process have to provide the required input information, such as transit and automobile time and cost figures between all zonal pairs for the target date, income and employment distribution for the same date, and whatever other parameters have been incorporated in the model. Figure 2.1 shows three hypothetical diversion curves, two representing trips to work from residential areas of low and medium income respectively, and one representing all non-work-related trips in an urban area.

The Abstract Mode Model

Models designed to estimate the demand for travel have various levels of specificity; they may estimate the number of trips to and from work during rush hour, or the number of commuter trips between residential suburbs and downtown, or the total number of trips between two traffic zones. The following model serves as an example for a research approach that simultaneously encompasses the objectives of trip generation, trip distribution, and modal split models. It also demonstrates that regression analysis can provide ample opportunity for creative research design and conceptual or methodological advancement.

Our example is the so-called abstract mode model, which estimates the number of trips T_{ijk} between urban areas i, j using a particular mode k. The estimation is of the linear regression type, expressing T_{ijk} as a function of socioeconomic characteristics of the areas i and j, and of selected characteristics of the mode k providing transportation between i and j. The idea of formulating trip numbers as a function of selected descriptors of the origin and destination—say, population, income, percentage of white-collar employment—comes straight from the familiar gravity model approach. It is the second set of variables—those describing the modes—and their representation in the model that deserve special attention.

A regression model based solely on the observation that certain phenomena covary—say, trip number with family size or income—has no explanatory value: Family members do not embark on trips because of the size of their family or the amount of family income. They travel in order to fulfill a desire or need or responsibility, whether it is buying groceries or serving on a jury or attending classes at a college.

Similarly, we may ask, why do individuals choose to travel by car or train, or why do they choose one over the other? Not because the mode of their choice is a bus or car, but because of the utility they associate with it: speed with airplanes, low cost with buses, flexibility with private cars. Thus, what matters is not the label of a mode but its performance characteristics, both absolute and relative to those of other modes competing with it. This is the reason for the concept of the "abstract" mode—a mode that is described solely by a set of numerical values measuring its performance.

For any mode k operating between locations i and j, the performance figures include the cost of a single trip from i to j, C_{ijk}, the associated time, H_{ijk}, and the frequency of departure, F_{ijk}. Individuals deciding on a particular mode usually arrive at that decision by comparing the different performance characteristics. To capture competition between modes the abstract mode model includes variables measuring relative performance: For each performance variable, the measure for mode k is divided by the measure of the mode with the best performance.

For example, if there are four modes (k = 1, 2, 3, 4)—say plane, bus, train, and car—and if their respective costs per trip from Los Angeles to San Francisco, are, respectively, $120, $60, $72, and $90, then the relative cost performance figures between these two cities would be 2.0, 1.0, 1.2, and 1.5 for the four modes, with C^b_{ij}, the best cost available between Los Angeles and San Francisco, being $60 and provided by the bus.

Individual decisions, however, are based not only on comparison—that is, on relative mode performances—but also on the absolute performance values. If each of the modal cost figures between Los Angeles and San Francisco were ten times as much, then the number of trips would decrease drastically, because many potential travelers either do not have the money or prefer to spend it on something else. It is sufficient to include in the equation estimating T_{ijk} the relative cost figure $C^r_{ijk} = C_{ijk}/C^b_{ij}$ of mode k and the best figure available, that is, C^b_{ij}, because the absolute cost for mode k, C_{ijk}, is included in the estimation by virtue of the relation $C_{ijk} = C^r_{ijk} \cdot C^b_{ij}$. Parallel considerations apply to other performance parameters, in our case, time and frequency of departure.

Altogether, the abstract mode model uses six mode performance measures for the estimation of the number of trips from i to j by mode k: the relative performance values for the time and cost and frequency of departure of the mode in question, and the best values available for any of the three performance parameters, irrespective of the modes by which they are provided.

At this point, the general form of the estimation equation is

$$T_{ijk} = f(P_i, P_j, I_i, I_j, H^b_{ij}, C^b_{ij}, F^b_{ij}, H^r_{ijk}, C^r_{ijk}, F^r_{ijk}). \qquad [14]$$

The first four independent variables measure population and income figures of origin and destination, i and j; the next three variables measure the best mode performances available—fastest time, least cost, most frequent departure between i and j; and the last three measure the particular performance of the mode in question, k, relative to the best performance figures among all competing modes. The functional form of the equation is assumed to be exponential, that is,

$$T_{ijk} = a \prod_{x=1}^{10} V_x^{b_x} \qquad [15]$$

where a is a constant, $V_x(x = 1, 2, \ldots, 10)$ are the ten variables listed above, and $b_x(x = 1, 2, \ldots, 10)$ are the exponents of the variables. The exponents b_x and the constant a have to be determined empirically. The symbol Π is equivalent to the symbol Σ, the only difference being that the different terms for $x = 1, 2, \ldots, 10$ are to be multiplied rather than added.

A logarithmic transformation will convert the equation into a linear form as needed for calibration by regression analysis. There is no a priori necessity that the independent variables V_x should have a multiplicative effect (rather than additive or any other) on the dependent variable T_{ijk}; however, the analysis of empirical data has shown that the functional form chosen for the model provides a reasonable description for the dependency observed.

In their original paper, Quandt and Baumol (1966) used the data of trips between sixteen pairs of California cities broken down by the modes of plane, bus, and private car. Trips by train had to be excluded because of insufficient data. As is unavoidable in this type of regression analysis, various sets of independent variables had to be examined before the set with the highest explanatory power (highest multiple correlation coefficient) could be determined. After repeated selection of variables and subsequent calibration, the following estimate was derived:

$$\begin{aligned} \log T_{ijk} = {} & -28.73 + 0.88 \log P_i + 0.88 \log P_j \\ & + 5.82 \log \frac{P_i Y_i + P_j Y_j}{P_i + P_j} \\ & -0.57 \log C^b_{ij} - 2.34 \log C^r_{ijk} \\ & -1.20 \log H^b_{ij} - 1.75 \log H^r_{ijk} + 0.44 \log F^r_{ijk}, \end{aligned} \qquad [16]$$

the multiple correlation coefficient being $R = 0.9386$. The sixth performance variable—the best frequency of departure—was omitted because it is always provided by the automobile and always has the same value, irrespective of origin, destination, or the departure frequencies of the other modes. For operational purposes, it was assumed that the private car departs every fifteen minutes (allowing for opening the garage, checking the car, starting the engine, and so on).

Let us study equation 16 in some detail. It has been obtained by first transforming all variables in equation 15 into their logarithms, thereby producing a linear relationship among them. Then the numerical values of the constants a and $b_x (x = 1, 2, \ldots, 10)$ were calculated from the sample of sixteen pairs of California cities. This was accomplished by means of regression analysis (see the first section of this chapter).

As a practical example, assume the trip origin and destination, i and j, to be Los Angeles (LA) and San Francisco (SF), and the mode of transport to be the airplane. Hence, T_{ijk} stands for the number of trips from LA to SF by air per unit time (for example, per work day). If we substitute the appropriate data for the independent variables on the right side of equation 16, it will produce a numerical estimate of the trip number T_{ijk}. We set

- P_i, P_j and Y_i, Y_j equal to the population and income figures for LA and SF.
- C^b_{ij} equal to the lowest cost available for a trip from LA to SF. The lowest cost is provided by the bus and is $60.
- C^r_{ijk} the relative airfare, equal to the actual airfare divided by the lowest cost available. If the actual airfare from LA to SF is $120, then the relative airfare is $120/$60 = 2.
- H^b_{ij} equal to 50 minutes, since this is the fastest travel time from LA to SF. The fastest time is, of course, by airplane.
- H^r_{ijk} equal to 1 because the airplane is itself the fastest mode.
- F^r_{ijk}, the relative frequency of airplane departure, equal to 32/96 = 0.33, where the numerator represents the number of plane departures from LA to SF per working day and the denominator is the daily frequency of departure by the mode with the highest frequency, that is, the automobile, which is assumed to be able to depart every 15 minutes.

At this point all numerical values of the parameters and variables of equation 16 are specified, and the equation will give us the logarithm of the numerical estimate of the number of trips from LA to SF by air per working day.

Several comments are in order:

(1) If we sum T_{ijk} over all modes k:

$$T_{ij} = \sum_k T_{ijk},$$

the abstract mode model becomes a distribution model providing estimates of total flow between two given locations. Likewise, summing over all destinations j and modes k transforms it into a trip generation model:

$$T_i = \sum_j \sum_k T_{ijk},$$

where T_i is the number of trips generated in i, for all modes and destinations.

(2) The abstract mode model not only will predict the flow volume of a particular mode for a given pair of traffic zones, but also will permit calculation of its growth or decrease if one of the mode's performance parameters is changed—for example, if bus fares go up by 10 percent or additional planes are introduced to increase air service. Moreover, the model will also predict what impact the performance change of one mode has on the others, and how the overall travel volume will change (but note the qualification of this statement below).

(3) The model will predict how much traffic a new mode will attract, even if this mode is completely different from all conventional modes and only its performance values are known. Whenever a new commercial mode is introduced, the model permits the choice of an appropriate price to maximize profit or to ensure that the capacity of the new mode is in line with the ridership it will attract.

(4) Not surprisingly, the model has its own set of shortcomings. Assume, for example, that in the case of California intercity travel the frequency of bus service were reduced. While the model would properly predict the decrease of bus passengers—the variable F^r_{ijk} in the equation estimating number of bus passengers would decrease—none of the equations estimating trip volumes of the other modes would be affected. This is simply the consequence of the model's design: Competition is expressed only by comparison with the mode that has the best performance and not by comparison with all modes. For example, the frequency performance of the plane is expressed relative to that of the car, which has the best performance for this characteristic; the change in bus frequency does not enter the estimating equation for plane travelers; hence, their number will be predicted as unchanged, in contrast to all empirical evidence, which tells us that if one mode reduces its frequency of departure it will usually lose some of its passengers to other modes.

(5) The signs of the variables in equation 16 correspond to our experience and expectation: Lower cost and faster connections, both relative and absolute, will increase trip volumes, and so will the relative increase of frequency of departure.

Although numerous improvements and extensions have been suggested in the literature, the abstract mode model has not become a standard planning tool; its data requirements are substantial, and its predictive capability has repeatedly been disappointing. However, it exemplifies the incremental progress of the travel demand research, that is, the increased abstraction of familiar concepts and introduction of new measures and techniques in the speculative modeling of processes underlying transportation decision making.

3. NETWORK LOADING

Suppose that we have arrived at an estimation of the volume, the origin, the destination, and the mode for each of a set of future traffic flows. How do we "load" these flows on existing or planned networks so as to be able to predict whether these networks are adequate? If they are not, how do we decide where and what kind of network improvements are needed? To answer these questions we first have to acquire a basic understanding of the theory of graphs and networks; this is the purpose of the following section.

Introduction to Graphs and Networks

In this section we will present concepts from graph and network theory which have proved indispensable in the description and analysis of transportation networks (Potts and Oliver 1972).

A graph is defined as a set V of elements called vertices (or nodes), and a set E of pairs of elements from V called edges (or arcs or links). Reflecting this definition, the notation of a graph is G(V,E), and it is visually represented by a set of points, certain pairs of which are interconnected by lines.

From this definition of a graph, it seems that we deal with an exceedingly simple, unstructured, trivial object—what could one possibly say about a cluster of points, some of which are connected by lines? For the uninitiated it is always an overwhelming experience to see how mathematicians have built formal structures of truly imposing architecture starting out with no more than a few primitive notions and relations—in this case, a set of objects, some of which form pairs (are "connected") with others. The theory of graphs is but one example, and within the limited scope of this text we will not be able to show more than the tip of the iceberg.

Graphlike structures permeate every facet of reality—the branches of a tree, the channels of a drainage system, the linkages among atoms in a molecule, but also the reporting structure in the army, the ties of friendship in a village, or the flow chart of a computer program. All of these phenomena can formally be described as graphs. Our concern here will be with the graph theoretical analysis of transportation networks.

Selected Definitions

Let G(V,E) be a graph. Each edge in E is defined by two vertices in V which are called the end points of the edge; the two vertices are said to be connected by the edge. Any two vertices connected by an edge are called adjacent; likewise, any two edges that have a common end point are called adjacent. The degree of a vertex is the number of edges which have that vertex as an end point, or, to phrase it differently, the degree of a vertex is the number of edges incident at that vertex. The sum of the degrees of all vertices in a graph is always twice the number of edges.

If the end points of an edge are identical, we call it a loop; different edges having the same end points are called parallel. Examples would be the two or more rail linkages between two adjacent cities built by competing railway companies. Graphs without loops or parallel edges are called linear. In this and subsequent chapters, we will restrict our discussion to linear graphs.

A set of pairwise different edges that form a sequence such that consecutive edges are adjacent in the graph is called a path; it is usually described by the sequence of vertices that define and interconnect the edges of the path. The two outermost vertices are called the end points of the path. If the end points are identical, then the path is called a cycle.

A graph can be represented by a drawing that displays the vertices and the interconnecting edges through points and lines. Alternatively, a graph can be described by a matrix—the so-called adjacency or connectivity matrix. The matrix has exactly one row and one column for each vertex of the graph, and the matrix cell defined by the row of one vertex and the column of another vertex contains a 1 or 0, depending on whether the two vertices are connected by an edge. Note that the matrix is symmetrical and that each row or column sum is equal to the degree of the corresponding vertex. Yet another complete description of a graph can be given in the form of two lists which identify explicitly all elements in the sets V and E. An airline timetable is an example: It lists all cities to which air service is provided, and it lists all direct flights, that is, the pairs of cities connected by nonstop flights.

Figure 3.1 shows the three different representations of a particular graph. The degrees of the vertices E and B are 3 and 2, respectively; the vertices A, E, C, D define a path between A and D, the vertices C, D, E, C define a cycle, and so do the vertices A, B, C, D, E, A.

If in a graph G(V,E) the set E consists of all possible pairs of elements in V, we call the graph complete. An example is provided by the ten

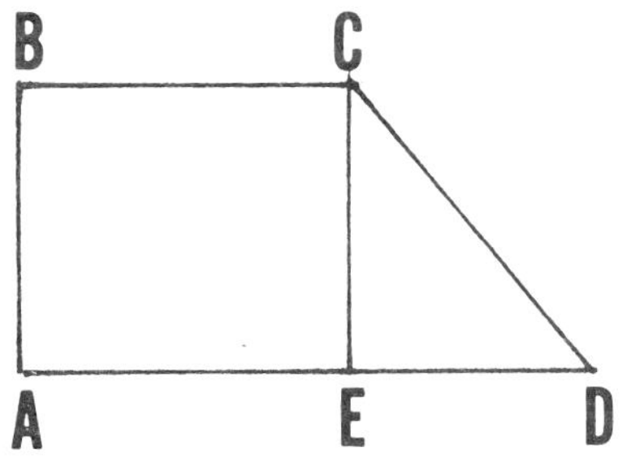

	A	B	C	D	E
A	0	1	0	0	1
B	1	0	1	0	0
C	0	1	0	1	1
D	0	0	1	0	1
E	1	0	1	1	0

$$V = \{A, B, C, D, E\} \qquad E = \{(A,B), (A,E), (B,C), (E,C), (E,D), (C,D)\}$$

Figure 3.1 Alternative Representations of a Graph

largest airports in the United States: Any two of them are interconnected by direct flights. We call G′(V′,E′) a subgraph of the graph G(V,E) if V′ is a subset of V and E′ contains all edges in E that join vertices of V′. If a graph can be drawn on a plane surface such that no edges intersect, it is called planar. The graphs of most ground transportation networks and almost all street networks fall under this category. A graph in which any two vertices are the end points of at least one path is called connected. A connected graph without a cycle is called a tree. Notice that the absence of any cycles in a tree is equivalent with the statement that any two vertices in a tree are connected by one and only one path. Examples of trees are natural drainage networks, the trees in a forest (which provided the name), or the pipes of a water distribution system.

The graph in Figure 3.1 is connected and planar. If we eliminate the edges (B,C) and (C,D), then the graph becomes a tree. The vertices E,C,D and the edges (E,C), (E,D), and (C,D) together form a subgraph that is complete.

There have been many research efforts to extract useful information from the graphs of real-world transportation networks (for a review, see

Leusmann 1979). As part of this research, a number of descriptive indices were introduced. An example is the beta index of a graph, defined as e/v, where e is the number of edges and v the number of vertices. There is no doubt that the index constitutes a measure of the connectedness of a graph. For example, the graphs of the railway networks in industrialized countries tend to have much higher beta values than those of Third World countries. Another index is the diameter of a (connected) graph: It is defined as the smallest number of edges by which any vertex can be reached from any other vertex, or, to state it more explicitly: If $l(x,y)$ is the number of links of the path with the smallest number of links connecting the vertices x and y of the graph G, then the diameter of G is the maximum of all values $l(x,y)$, where x and y represent all possible pairs of vertices in G.

Other indices describe the position of individual vertices in a graph. For example, a measure of accessibility of a vertex x would be the minimum number of edges it takes to reach any one vertex y from x (the so-called associated number of the vertex x); a more comprehensive measure would be the sum of all values $l(x,y)$, where x is again the vertex whose accessibility we wish to measure and y represents each of the other vertices in G.

The beta index of the graph in Figure 3.1 is 1.2; the diameter is 2, and all vertices have the same associated number, namely 2; if, however, we would eliminate the edge (A,E), then the associated numbers of the vertices A,E and D would become 3.

If we know more about a graph than its adjacency matrix, and if this additional knowledge is used for description or analysis, then we will call such a graph a network. In the case of transportation networks, this additional information typically includes the length and capacity of each edge (now called a link) and the capacity of each vertex (now called a node); however, depending on the purpose of a particular network analysis, it might also contain data on construction cost, accident rate, or congestion for some or all links and nodes of the network.

As a first example of network analysis, we will describe and solve a tranportation problem well known as "the maximum flow through a network." This problem is of applied interest not only in wartime or during other emergencies, but also, more generally, in all those fairly frequent situations when the existing demand for transportation requires maximum utilization of the unused capacity of a network. Even more important is another problem in transportation planning: how to find the shortest path through a network. We will first demonstrate that the second problem is equivalent to the first and will then solve the second problem (thanks to the equivalence, the solution of one problem will solve the other problem as well).

The Maximum Flow-Minimum Cut Theorem

Let G(V,E) be the graph of a transportation network. We denote the capacity available on the link (i,j) by c(i,j), where capacity is measured in units of flow per unit time. Let A and B be two nodes in G. What is the maximum possible amount of flow from A to B per unit time? Figure 3.2 provides an example of the problem.

Obviously, the network contains a number of different paths by which flows can be moved from A to B. Notice, however, that the maximum flow cannot be more than fifteen units, since this the maximum flow volume that can be shipped out of A via the three links connecting A with the rest of the network. While the capacities available elsewhere in the network cannot increase this limit of fifteen units, they can certainly decrease it—just check the capacities of the links incident at B: Their combined capacity is only twelve, and this figure is now our new upper limit of the maximum possible flow from A to B. Can the remaining network handle the flow of twelve units? To answer this question systematically, we introduce the concept of a cut.

A cut C(A,B) in a network G is defined as a set of links in G with the following two properties:

(1) If the links of the cut are removed from G, then there is no path in G connecting A and B.
(2) If fewer than all of the links of the cut C(A,B) are removed from G, then G contains at least one path connecting A and B.

The capacity of a cut is defined as the sum of the capacities of the links in the cut.

In Figure 3.2, the links (H,G), (E,F), (E,D), and (C,D) form a cut with regard to the nodes A and B, and so do the links (H,G), (G,F), and (D,B). The two sets of links we considered earlier—those from A to H,E and C as well as the links from B to G and D—also constitute cuts with regard to A and B.

Notice that all shipments from A to B have to move through all cuts separating A and B. Evidently, the maximum possible flow from A to B cannot be more than the minimum value M of the capacities of all cuts separating A and B. On the other hand, it can be at least that much, because, if there were a combination of links in G that would keep the flow volume below M, then these links would constitute a cut with a cut capacity less than M, in contradiction to our assumption that M is the minimum cut capacity with regard to A and B. Thus, the maximum possible flow from A to B is equal to the minimum capacity of all cuts separating A and B—hence the abbreviated name "max-flow/min-cut theorem." Going back to Figure 3.2, the cut with the smallest capacity is

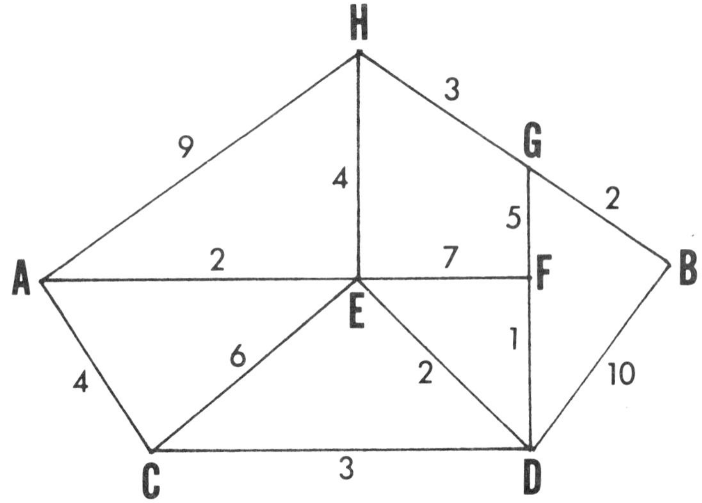

Figure 3.2 Transportation Network with Link Capacities

the set of links (C,D), (E,D), (D,F), (G,B). The capacity associated with this cut is 3 + 2 + 1 + 2 = 8, which is therefore the maximum flow that can be shipped from A to B.

We now transform the maximum flow problem into one of finding the shortest path through a network. Let G be a planar graph. The areas that are enclosed by cycles and do not contain any other cycles are called faces, and the area outside the cycles of G is called the infinite face. Thus, the faces together with the infinite face of a planar graph produce an exhaustive subdivision of the entire plane into mutually exclusive areas.

We construct a new graph, G′, as follows:

(1) We place a node of G′ in each face of G, including its infinite face.
(2) If x and y are two nodes of G′ whose corresponding faces in G share a link, then we connect x and y by a link.

If we were to start out with the graph G′, we would get G as the new graph; it is because of this peculiar relationship that the graphs G and G′ are called dual graphs.

Let G be a planar network with given link capacities. We wish to determine the maximum possible flow from node A to node B in G. To this end, we add a new link (A,B) to G, which connects A and B and has zero capacity. We then construct the dual network G′ of the enlarged network G. Let A′ and B′ be the two nodes in G′ which correspond to the two faces of G having the link (A, B) in common. Notice that each link e in G is intersected by one and only one link e′ of G′. To each link e′ of G′ we assign a length which is equal to the capacity value of the corresponding link e in G. Then the following statements hold:

I. Every path from A′ to B′ intersects a set of links in G that form a cut separating A and B in G.
II. The set of links in G′ that correspond to the links in a cut C(A,B) in G form a path connecting A′ and B′ in G′.
III. The length of a path connecting A′ and B′ in G′, if defined as the sum of the lengths of its links, is equal to the capacity of the cut in G defined by this path.
IV. Finding the maximum flow of A to B in G is equivalent to the problem of finding the shortest path between A′ and B′ in G′, and the length of the shortest path from A′ to B′ is equal to the maximum flow from A to B.

Figure 3.3 shows the graph G of Figure 3.2 with the link (A, B) added, as well as the dual graph G′.

Having transformed the maximum flow problem into that of finding the shortest path in the dual network, we will now develop a solution procedure that will identify the path of minimum length (or time) for any pair of nodes in a given network.

Trip Assignment

Let us assume that, in the course of preparing a transportation plan, a trip distribution model has given us a complete origin/destination table of traffic flows to be expected for some future date. To decide whether the current transportation network will be adequate to handle those future flows, we simulate its performance for the new flow pattern.

The process by which a transportation planner assigns individual flows to individual network paths to simulate an expected network flow pattern is called network loading or trip assignment. In a strictly planned economy, the movement of people and freight can be channeled in some optimal fashion, linear programming providing one approach that will determine the optimal assignment of individual flows to network paths under a variety of constraints and objectives. Examples of constraints are capacity limits on individual links, and examples of objectives are the maximization of average traffic velocity or the minimization of total vehicle miles.

However, if individuals decide about their trip-making behavior themselves, and if the network in question is the road network of a

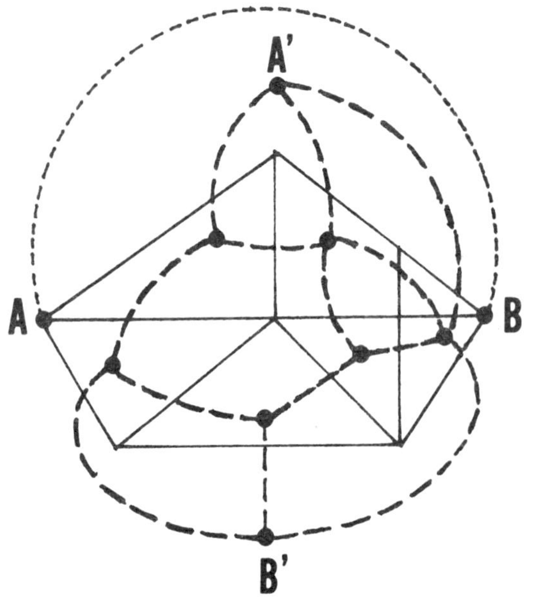

Figure 3.3 A Transportation Network and Its Dual Network

metropolitan area, then the task of predicting the pattern of paths travelers will choose on that network is marred with difficulties. While it has a ring of plausibility, empirical evidence does not support the hypothesis that travelers always choose the shortest path for their trip. Nonetheless, it is the basis for most network-loading techniques and is known under the name "all or nothing assignment." We will discuss selected improvements and refinements later on, but first we present an algorithm which will generate the shortest path between any two network nodes.

The Shortest Path Through a Network (Dynamic Programming)

Dynamic programming is a systematic technique of optimizing a sequence of decisions in which the optimality of the final outcome depends on the optimality of each individual decision. Finding the optimal (shortest, fastest) path through a network is a good example: At each node we reach, we have to decide on the next node, and we will not reach the destination node over the optimal path unless the decisions on all previously selected nodes were optimal.

Before solving the problem of finding the shortest path between two nodes A_1, A_n in a given network, we will first solve the problem of the shortest path from A_1 to A_n in k steps, where each step constitutes a decision to include a particular node in the path. We will not require that a node can be included only once; thus, a path from A_1 to A_n generated in k steps has at most k links. Once we know the shortest paths from A_i to A_n for all values of k, then the shortest path in this set is necessarily identical to the shortest path from A_1 to A_n.

The shortest path from any node to some other node will at most include all network nodes. Therefore, if the number of nodes in the network is m, then the shortest path between two nodes can be generated in at most m − 1 steps.

Notation: Let $L(A_i)$ be the length of the shortest path from A_i to A_n; let $L_k(A_i)$ be the length of the shortest path from A_i to A_n in k steps; and let $L(A_i, A_j)$ be the length of the link from A_i to A_j. In particular, we will define this length to be equal to 0 if $i = j$, and we will define it to be infinity if there is no direct link between the nodes A_i and A_j.

A path P from some network node A_i to A_n in k steps ($k > 0$) will connect A_i with some node A_j, and will connect A_j with A_n in $k - 1$ steps. Given the way we have defined what constitutes a step, both $i = j$ and $j = n$ are possible. Notice that P will be the shortest path from A_i to A_j to A_n in k steps only if the segment from A_j to A_n is the shortest path between these two nodes in $k - 1$ steps. The length of P is therefore $L(A_i, A_j) + L_{k-1}(A_j)$. However, this condition is not sufficient, as our choice of A_j was arbitrary: It might very well be that connecting A_1 with some other node, say A_x, and then continuing from A_x in $k - 1$ steps to A_n might provide a shorter connection from A_i to A_n in k steps. Apparently there is a total of m choices of nodes to select as A_x, because A_x can be any one of the m nodes in the network. The optimal choice will be the node which provides a connection that is of minimum length:

$$L_k(A_i) = \mathrm{Min}\left\{L(A_i, A_x) + L_{k-1}(A_x) \mid x = 1, \ldots, m\right\} \qquad i = 1, \ldots, m. \qquad [17]$$

This last equation provides the key for our overall problem because, for every node A_i in the network, it expresses the shortest path to A_n in k steps as a function of the lengths of network links (which are known) and of the shortest paths from all network nodes to A_n in $k - 1$ steps. Thus, we have a recursive relationship allowing us to calculate the shortest path from any node to A_n for any number of steps.

We start the recursive process with $k = 1$; that is, we find the shortest path from each node to A_n in one step. To reach A_n from A_i in one step requires that the path, in addition to the node of origin, A_i, has at most one more node, which must therefore be the destination node A_n. Thus, for each A_i, $i = 1, 2, \ldots, m$, there is only one choice of node to pick next, namely A_n, and the length of the shortest path from A_i to A_n in one step is therefore

$$L_1(A_i) = L(A_i, A_n) \quad \text{where} \quad i = 1, 2, \ldots, m. \qquad [18]$$

For $k = 2$ we get, according to our recursive equation,

$$L_2(A_i) = \text{Min}\{L(A_i, A_x) + L_1(A_x) \mid x = 1, \ldots, m\}$$

$$= \text{Min}\{L(A_i, A_x) + L(A_x, A_n) \mid x = 1, \ldots, m\} \qquad [19]$$

$$\text{where} \quad i = 1, \ldots, m.$$

For $k = 3$ it is

$$L_3(A_i) = \text{Min}\{L(A_i, A_x) + L_2(A_x) \mid x = 1, \ldots, m\} \qquad [20]$$

$$\text{where} \quad i = 1, \ldots, m$$

and so on. Each $L_k(A_i)$, can be readily calculated using the set $L_{k-1}(A_i)$ and the length values of the network links (equation 17). It is important that we record, at each step, which particular node A_x minimizes equation 17 so that we can ultimately describe each shortest path by the sequence of nodes that define its links. Once we have successively determined the shortest paths from all network nodes to A_n in $k = 1, 2, 3, \ldots, m - 1$ steps, the length of the shortest path from A_1 to A_n is simply

$$L(A_1) = \text{Min}\{L_k(A_1) \mid k = 1, \ldots, m - 1\}. \qquad [21]$$

Example: The network in Figure 3.4 consists of five nodes and seven links, the length of each being indicated by a number (say, in miles) next to it. We wish to determine the shortest path from A_1 to A_5. Recall that the length of the direct connection of a node to itself is 0, while the length of a direct connection between two nodes not connected by a link is defined as infinity. In line with the notation introduced earlier, $L_k(A_i)$ is the length of the shortest path from A_i to A_5 in k steps, where $k = 1, 2, 3, 4$.

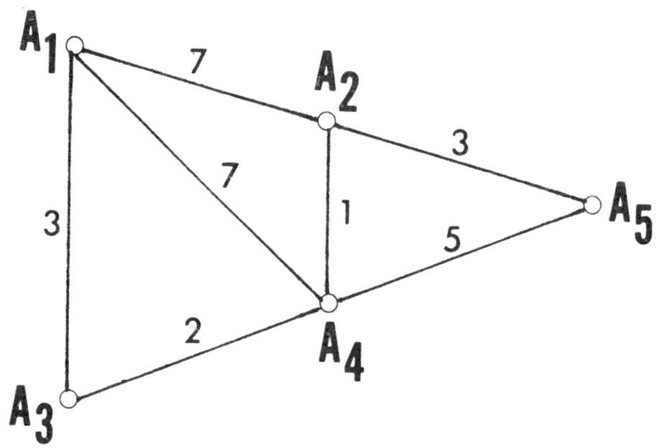

Figure 3.4 Network with Specified Link Lengths

Evidently, it is $L_1(A_i) = \infty, 3, \infty, 5, 0$ for $i = 1, 2, 3, 4, 5$. To demonstrate the application of the recursive process, we will write out in detail the calculations of $L_2(A_1)$, that is, the shortest route from A_1 to A_5 in two steps:

$$L_2(A_1) = \text{Min} \begin{Bmatrix} L(A_1, A_1) + L_1(A_1) = 0 + \infty = \infty \\ L(A_1, A_2) + L_1(A_2) = 7 + 3 = 10 \\ L(A_1, A_3) + L_1(A_3) = 3 + \infty = \infty \\ L(A_1, A_4) + L_1(A_4) = 7 + 5 = 12 \\ L(A_1, A_5) + L_1(A_5) = \infty + 0 = \infty \end{Bmatrix}$$

or $L_2(A_1) = 10$. Likewise,

$$L_2(A_2) = \text{Min}\left\{L(A_2, A_i) + L_1(A_i) \mid i = 1, 2, 3, 4, 5\right\}$$
$$= \text{Min}\left\{7 + \infty,\ 0 + 3,\ \infty + \infty,\ 1 + 5,\ 3 + 0\right\} = 3.$$

The complete set of length values of the shortest paths from any node A_i to A_5 in k steps ($k = 1, 2, 3, 4$) is provided in Table 3.1. (Underneath

TABLE 3.1 **Shortest Paths from Each Node of the Network of Figure 3.4 to the Node A_5 in k Steps, where k = 1, 2, 3, 4**

$A_i \rightarrow$	A_1	A_2	A_3	A_4	A_5
$L_1(A_i)$	∞	3	∞	5	0
	A_5	A_5	A_5	A_5	A_5
$L_2(A_i)$	10	3	7	4	0
	A_2	A_2 or A_5	A_4	A_2	A_5
$L_3(A_i)$	10	3	6	4	0
	A_1 or A_3 or A_4	A_2 or A_5	A_4	A_2 or A_4	A_5
$L_4(A_i)$	9	3	6	4	0
	A_3	A_2 or A_5	A_2 or A_4	A_2 or A_4	A_5

each value you will find the node that has been included in the shortest path in that particular step.)

Table 3.1 provides the answer to the original question: What is the shortest path from A_1 to A_5? The length of that path is

$$L(A_1) = \text{Min}\left\{L_k(A_1) \mid k = 1, 2, 3, 4\right\} = \text{Min}\left\{\infty, 10, 10, 9\right\} = 9.$$

The table also permits us to reconstruct the shortest path from A_1 to A_5 as follows:

We start with the cell in the A_1 column containing the shortest path length from A_1 to A_5. This is the fourth cell, indicating that the path consists of four steps. The cell also specifies the node in the path adjacent to A_1, namely, A_3.

From A_3 the path will consist of three steps, and we therefore turn to the cell in the A_3 column and the third row. This cell tells us that the remaining length of the path is 6 and that it proceeds via A_4.

From A_4 there are two steps left, and we therefore inspect the cell in tha A_4 column and the second row. The information in this cell indicates that the remaining length of the path is 4 and that it continues via A_2.

With one step of the path left, we turn to the cell in the A_2 column and the first row. It indicates that the remaining path length is 3 and that the next node is A_5, that is, the path has reached its destination. Summarizing, the path consists of the sequence of nodes A_1, A_3, A_4, A_2, A_5.

The following observations are noteworthy:

(1) The solution process as presented here provides us with the shortest route not only from a particular origin node to a destination node but also from all nodes in the network to that particular destination node.

(2) If the network is not directed—that is, if flows can take place in both directions of each link and the length of each link is the same in both directions—then we can start the recursive process from the other end and will get not only the shortest path from A_n to A_1 (which is the shortest path from A_1 to A_n) but also the shortest path from any node in the network to A_1.

(3) It is only a matter of our particular choice that we have calculated the shortest path, where the length of a path is defined as the sum of the length of its links. Without any change in the algorithm, we can also determine the minimum cost path whereby each link has a particular cost associated with it and the cost of a path is defined as the sum of the individual link cost. These costs could be, for example, toll charges or fuel expenses. Similarly, the algorithm will determine the fastest path if we define the time of a network path by the sum of the times it takes to traverse the individual links of that path.

(4) If the shortest path from A_1 to A_n passes through the nodes A_x, A_y, then the segment of the path connecting A_x with A_y is identical to the shortest path from A_x to A_y.

(5) It is self-evident that the shortest path between two nodes need not be the path with the smallest number of links. It is also clear that the shortest path will not contain any cycles; that is, it passes through each network node at most one time.

4. OPTIMAL TRANSPORTATION DECISIONS THROUGH LINEAR PROGRAMMING

Concepts of Linear Programming

We will introduce the purpose and the main components of a linear program by way of an example.

Consider a mining company shipping coal from its mine to a central distribution point by barge and train. The following statements describe the conditions under which the company operates, and they formulate questions regarding profit maximization and optimal decision making.

(1) The production capacity of the mine is 9,000 tons per unit time.

(2) Delivery contracts require the company to ship at least 4000 tons per unit time.
(3) The available shipping capacity of the two transportation modes is 5500 tons each, again per unit time.
(4) There are 49 pieces of loading equipment available, 7 of which are needed for each 1000 tons of coal loaded onto barges, and 3 for each 1000 tons loaded onto railroad cars.
(5) The loading of 1000 tons of coal on barges requires 14 units of labor, and the same amount loaded on railroad cars requires 42 units of labor. Agreements with the labor unions stipulate the employment of at least 147 units of labor.
(6) For each unit of time, contractual obligations with the transportation companies call for at least 1500 tons to be shipped by barge and 1000 tons by train.
(7) Finally, the profit on 1000 tons of coal shipped by barge is $700, and $1400 if shipped by train.

These statements do not constitute, by themselves, a problem, but simply represent a quantified description of a particular and highly simplified economic scenario. Mining company managers and stockholders have a number of goals and concerns that translate into specific questions, such as the following:

I. Given the circumstances described above, what is the maximum profit possible?
II. How much coal should be produced, and how much should be shipped by barge versus train?
III. Of the economic conditions itemized above, which ones prevent profits from being higher?

Note: The problem as stated here is already in a form that permits its translation into mathematical language without any difficulty. For many real-world problems, the most intractable step in the solution process is frequently not the lack of solution methods but rather the difficulty of formulating the problem so as to make it amenable to the application of existing methods.

Either explicitly or implicitly, both the descriptive statements and the questions listed above refer to the volume of shipment by the two modes. Let x be the amount of coal shipped by barge, and y the amount shipped by train, both measured in units of 1000 tons. Then the individual statements 1 through 7 describing the conditions or "constraints" under which the mining company has to operate can be algebraically expressed in terms of x and y as follows:

(1) $x + y \leq 9$; that is, the combined shipment by barge and train must not exceed the production capacity of the mine (as always, per unit time).

(2) $x + y \geq 4$; that is, the combined shipment has to be at least 4000 tons.
(3) $x \leq 5.5$ and $y \leq 5.5$; that is, the shipments by each mode must not exceed the transportation capacity available on these modes.
(4) $7x + 3y \leq 49$; this condition follows from the following consideration: If the tonnage shipped by barge is x and if it takes 7 pieces of loading equipment loading a thousand tons on barges, then the number of equipment pieces required is $7 \cdot x$; likewise, it takes $3 \cdot y$ pieces of equipment to load y tons onto railroad cars, where x and y are again measured in thousands of tons. Thus, loading x and y tons on barges and trains, respectively, requires $7x + 3y$ pieces of equipment. Since there are only 49 available, any solution specifying numerical values for x and y has to satisfy the condition $7x + 3y \leq 49$.
(5) Parallel to the reasoning in statement 4, the shipment of x and y tons of coal by barge and train requires 14x and 42y units of labor, respectively; since at least 147 units of labor have to be employed, the following condition must be met: $14x + 42y \geq 147$.
(6) These conditions translate into $x \geq 1.5$ and $y \geq 1.0$.
(7) The profit P is a linear function of the amounts of coal shipped by barge and train: $P = 700x + 1400y$, as always, per unit time

We can now state our problem as follows: Find numerical values for the unknowns x,y, such that the profit P is maximized and all conditions are met:

$$\text{Max } P = 700x + 1400y \qquad [22]$$

s.t. (subject to):

(1) $x + y \leqslant 9$	(4) $7x + 3y \leqslant 49$
(2) $x + y \geqslant 4$	(5) $14x + 42y \geqslant 147$
(3a) $x \leqslant 5.5$	(6a) $x \geqslant 1.5$
(3b) $y \leqslant 5.5$	(6b) $y \geqslant 1.0$

The two variables x,y are called the decision variables; equation 22 is called the linear objective function (it is linear in the decision variables and specifies the objective, that is, maximization of profit); the conditions listed with equation 22 are called linear constraints, obviously because they limit the values that are permissible for the decision variables and because they are linear in these variables. The linear objective function together with the set of linear constraints is called a "linear program."

Next, let us construct the geometric figure corresponding to the set of constraints. We recall from high school algebra that each of the inequalities representing a constraint defines a half plane in the two-dimensional plane. Take, for example, constraint 3a: $x \leq 5.5$. Let the two-dimensional plane be defined by a system of x,y coordinates. Then each point in the plane is specified by its x and y coordinates. Consider all points for which the first coordinate, x, is smaller than or equal to 5.5. Apparently all points qualify as long as they lie on, or to the left of, the straight line $x = 5.5$. This line runs parallel to the y-axis and intersects with the x-axis in the point (5.5;0). Similarly, all points x,y lying on or above the line $14x + 42y = 147$ fulfill constraint 5. We can reformulate this latter equation so as to give it a more familiar form,

$$y = \frac{7}{2} - \frac{1}{3}x,$$

and represent it graphically in the x,y plane as a straight line.

Figure 4.1 shows all eight constraints by the half planes they define. Each half plane is indicated by its limiting line (corresponding to the equality part in the inequality), with arrows specifying the side on which it is located. The linearity of the constraints shows up graphically in the limiting lines being "straight" rather than curved.

The solution we are after is a pair of x,y values that maximizes the objective function and meets all constraints. Thus, the solution (if there is one) will be a point which is located in all eight half planes, that is, in their intersection. This intersection is the (shaded) polygon ABCDE: All points inside meet all constraints, and all points outside violate one or more. It is called the solution space, and each point in it is called a feasible solution—feasible because it satisfies all constraints. A feasible solution point is not necessarily optimal, but an optimal solution point has to be feasible, which is to say it has to be located in the solution space.

Now that we have nicely delineated the area of all feasible solutions, how do we find the one that is optimal, that is, the pair of x,y values that maximizes the objective function $P = 700x + 1400y$? This function cannot readily be displayed in our figure because the figure is two-dimensional, in x and y, while the profit function is three-dimensional, with variables P, x, y. We get around this problem by simply choosing a numerical value for P, say, $2800. Thus, $700x + 1400y = 2800$, or $y = 2 - 0.5x$. Every pair of numerical values satisfying this equation will render a profit of $2800, at least if, for the moment, we ignore all constraints. Since the equation is stated in the two dimensions of our system of coordinates, we can readily represent it in the figure as a straight line (line P_0). Notice that this line does not intersect with the solution space, which means that all combinations of barge/train ship-

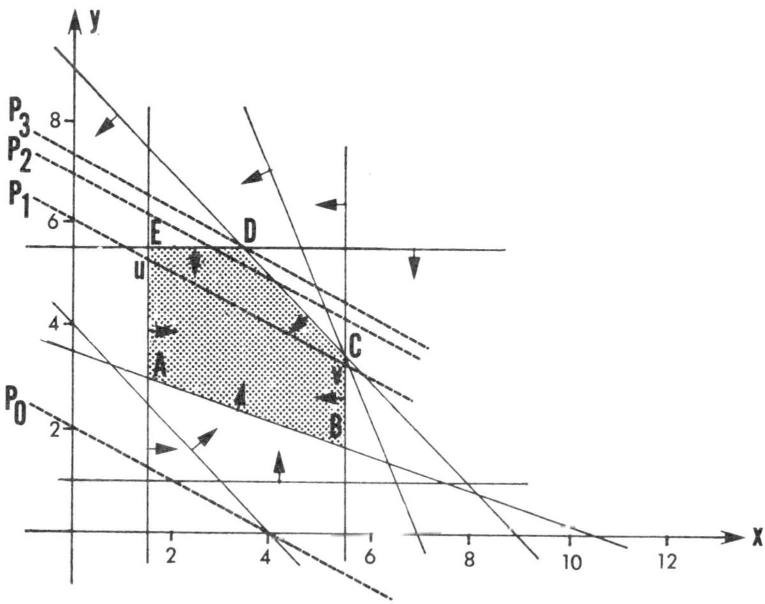

Figure 4.1 Graphical Solution of a Linear Program in Two Variables

ments x,y producing a profit of $2800 violate at least one of the constraints.

Before we continue, let us rewrite the profit function 22 in terms of y:

$$y = \frac{P}{1,400} - \frac{1}{2}x$$

Note that for each numerical profit value we will get a particular straight line in the graph; note also that changing the profit will change the constant (the "absolute value") in the equation but not the coefficient of the independent variable, x. Thus, the family of curves defined by different profit values P consists of a set of straight lines, all having the same slope ($-\frac{1}{2}$); they are therefore parallel.

Let us now, as a second trial, triple the profit and make it $8400. As we have just established, the corresponding line P_1 in Figure 4.1 is parallel to P_0. More important, it passes through the solution space: All points on the line between U and V are feasible solutions of barge and train shipments, all generating the same profit of $8400. Is it possible to obtain a still higher profit? We rephrase the question in terms of our figure: Are there any points x,y in the solution space that generate a higher profit, that is, points that satisfy the condition $700x + 1400y > 8400$? The answer is obviously yes: All points located both in the solution space and in the half plane defined by the inequality $700x + 1400y > 8400$ meet these two requirements. They are all and only the points located in the subpolygon UVCDE, excluding the points on the boundary between U and V.

Increasing the profit from 8400 to a higher value means shifting the profit line further away from the origin while maintaining its slope: The line P_2 provides higher profits while still intersecting with the solution space. The maximum profit solution is apparently reached when we move the line until it touches only the solution space (line P_3) and any further shift would render the intersection between the profit line and the solution space empty. This is obviously the case at point D, which therefore constitutes the maximum profit solution for our problem: $x = 3.5$ and $y = 5.5$.

We are now in the position to answer the questions raised earlier:

I. The maximum possible profit per unit time is $700 \cdot 3.5 + 1400 \cdot 5.5 =$ $10,150.
II. Profit will be maximized when the total shipment is 9000 tons, of which 3500 are shipped by barge and 5500 by train.
III. The constraints that prevent the profit line from moving further are those of the production capacity of the mine and the transportation capacity of rail transport. These are the constraints effectively restricting the profit from being higher and should therefore be given priority in investment decisions geared toward improving profits.

Our example, with its graphical solution, reveals several other features about the nature of a linear program. For one, unlike the maximal or minimal points of functions determined by calculus, the optimal solution points of a linear program are always located on the boundary of the solution space rather than on its inside; specifically, with the exception of particular circumstances (one of which we discuss below), the optimal solution point coincides with one of the vertices of the polygon delineating the solution space.

It also becomes clear that a linear program need not have just one solution. If, in our example, profit per thousand tons would be equal for shipment by barge and by train, then the last profit line intersecting with the solution space would in fact be identical to the line representing the mining capacity constraint, and all points on the segment C,D would yield the same maximum profit.

If, on the other hand, there were no mining capacity constraint and the shipping capacity of either barge or train were removed, then the polygon defining the solution space would be open and the profit line could be moved to the right indefinitely without leaving the solution space. In this case the mathematical solution would be $x,y = (5.5;\infty)$ or $(\infty;5.5)$; in reality, of course, there is no such thing as unlimited transport capacity.

Finally, there is always the possibility that the intersection of the half planes defined by the constraints is empty. If, in our example, we replace the barge capacity constraint $x \leq 5.5$ with $x \leq 1$, then no feasible solution would exist, because any choice of shipping quantities x,y would be in violation of one or more constraints.

Notice that of the eight constraints, only five participate in the definition of the solution space, the other constraints being redundant.

There are two constraints that we did not even introduce because of redundancy, although they are always part of a standard linear program: These are the conditions that all decision variables are non-negative—in our case, $x \geq 0$ and $y \geq 0$. Since we already have the "stronger" constraints $x \geq 1.5$ and $y \geq 1.0$, the former constraints are rendered redundant. Notice also that the $y \geq 1.0$ constraint is, in turn, dominated by others, so that even it does not participate in the delineation of the solution space.

The Hitchcock Linear Programming Model and Its Solution

A simple but practical example of a linear programming application to transportation flow planning has been provided by Maurice Yeates (1963) in his normative study on high school hinterland delimitation in southwestern Wisconsin. Formally, it is identical to the classical Hitchcock problem (Hitchcock 1941).

Consider a set of m high schools labeled $j = 1, 2, \ldots, m$, with capacities C_j and a set of n areas labeled $i = 1, 2, \ldots, n$ with student populations P_i where the total student population is equal to the total available high school capacity:

$$\sum_{i=1}^{n} P_i = \sum_{j=1}^{m} C_j. \qquad [23]$$

Let d_{ij} denote the distance between area i and high school j, measured in road miles. Question: How many students x_{ij} residing in area i should be assigned to the high school j where i and j assume all possible origins and destinations, such that the total daily student mileage L is minimized?

Evidently, the objective function is

$$\text{Min } L = \sum_{i=1}^{n} \sum_{j=1}^{m} x_{ij} d_{ij} \qquad [24]$$

and is subject to two sets of constraints. They guarantee that the number of students from area i assigned to the various high schools is at least as large as the number P_i of students residing in i, where i may be any of the n areas:

$$\sum_{j=1}^{m} x_{ij} \geqslant P_i \qquad i = 1, \ldots, n. \qquad [25]$$

The second set of constraints guarantees that no high school is overloaded:

$$\sum_{i=1}^{n} x_{ij} \leqslant C_j \qquad j = 1, \ldots, m. \qquad [26]$$

To prevent the linear program from generating a solution containing negative flows, an additional set of constraints stipulates that the flows x_{ij} satisfy $x_{ij} \geq 0$ for all pairs i,j. The numerical values of the student flows x_{ij} of the minimum cost solution identify, for each school, the area around it from which it should draw its student population, with some areas sending students to more than one high school; clearly, these latter areas lie on the boundaries of the school commuter hinterlands.

The complexity of the problem, and thereby its usefulness, can easily be increased by adding further decision variables and constraints. If, for example, transport capacities are limited, then a new set of constraints $x_{ij} \leq C_{ij}$ can readily be incorporated into the linear program, where C_{ij} denotes the available transport capacity between i and j, and i = 1, . . . , n; j = 1, . . . , m.

An additional and more interesting extension would be the following: Let Q_i(i = 1, . . . , n) be the student population in area i as projected for some future time, say the year 2000; let k_j be the annual cost for each additional study place added to the capacity of high school j, j = 1, . . . , m; let t be the cost per student mile per year; and let y_j be the (as yet unknown) amount of capacity to be added to high school j so as to meet the expected increase in student enrollment. Problem: Determine both the capacity increase at each of the m high schools and the assignments of students to high schools such that the combined cost of capacity addition and student transportation is minimized:

$$\text{Min } C = t \cdot \sum_i \sum_j x_{ij} d_{ij} + \sum_j y_j k_j \qquad [27]$$

s.t.:

$$
\begin{aligned}
&\sum_j x_{ij} \geqslant Q_i && i = 1, \ldots, n \\
&\sum_i x_{ij} - y_j \leqslant C_j && j = 1, \ldots, m \\
&x_{ij} \geqslant 0 \quad y_j \geqslant 0 && \\
& && i = 1, \ldots, n;\ j = 1, \ldots, m. \\
&x_{ij} \leqslant C_{ij} &&
\end{aligned} \qquad [28]
$$

As formulated, the problem allows for the building of new high schools: We introduce them with the labels m + 1, m + 2, . . . and existing capacities $C_{m+1} = C_{m+2} = \ldots = 0$ and added capacities $y_{m+1}, y_{m+2}, \ldots$. Should the minimum cost solution assign capacity figures to some of the new high schools too small to warrant the effort, they can simply be eliminated from the set of all schools and the linear program can be run again.

In the simple form presented above, the Hitchcock problem can be solved by a simple iterative process known as the "stepping stone procedure" (Charnes and Cooper 1958). To demonstrate the principles of this procedure, we return to the original formulation of the student assignment problem.

Given that the total number of students and the sum total of school capacities in this example are assumed to be equal, constraints 25 and 26 imposed on the student flows leaving an area i or arriving at a high school j have to be equalities. Let us arrange the flows x_{ij} in matrix format such that each row corresponds to an area i and each column to a high school j. Then the sum of row i constitutes the outflow from the area i and must be equal to its student population P_i, and the sum of column j forms the inflow into high school j and must therefore be equal to its capacity C_j.

The so-called "northeast corner rule" provides a first feasible solution. Starting with the first row, we move from cell to cell along each row, placing into each cell always the maximum figure that will not violate either the row sum or the column sum reduced by the amount of flows already allocated in previous steps. Notice that with each step either a row sum or a column sum will be balanced, and that the remaining cells in that row or column will therefore receive zeroes. Notice also that, as a consequence, the number of positive flows will at most be n + m − 1.

Comparing the feasible solution of student flows generated by the northwest corner rule with the distance matrix $\{d_{ij}\}$ will in most cases quickly reveal that this feasible solution is not optimal; that is to say, there are other feasible solutions producing a lower total student mileage figure. To generate a new feasible solution, we select four flows from the first feasible solution that together form a rectangle, as shown in Table 4.1. Next, we change their values by an amount f, where f is as large as possible without reducing any flow below zero. Since the row and column sums remain balanced, the result of this flow transfer is a new feasible solution, and it is an improved solution if the mileage figures associated with the increased flows are less than those associated with the reduced flows. In the terms of Table 4.1, this condition translates into $d_{ij} + d_{uk} < d_{uj} + d_{ik}$. This method is repeatedly applied until no further improvement is possible, indicating that the optimal solution has been reached.

Example: Let $C_1 = 80$ and $C_2 = 120$ be the capacities of two coal distribution centers, and let $D_1 = 20$, $D_2 = 70$, and $D_3 = 110$ be the demand figures of three cities 1, 2, 3, where capacities, demands, and flows are measured in thousands of tons. Let the distances between depot 1 and the three cities be, respectively, 30, 20, and 15 miles, and between depot 2 and the cities 5, 10, and 15 miles. What are the numerical values of the six flows that will minimize the total ton mileage while satisfying all capacity and demand constraints?

The northwest corner rule gives us a first feasible solution:

$$x_{11} = 20 \qquad x_{12} = 60 \qquad x_{13} = 0$$

$$x_{21} = 0 \qquad x_{22} = 10 \qquad x_{23} = 110 \qquad \text{Total ton mileage} = 3550.$$

First improvement: Shift 10 flow units from x_{11} to x_{21} and from x_{22} to x_{12}. Total ton mileage: 3400. Second improvement: Reallocate 10 units between x_{11}, x_{21}, x_{23}, and x_{13}. Third improvement: Reallocate 70 units between x_{12}, x_{22}, x_{23}, and x_{13}. At this point no further improvement is possible, and the last solution is therefore optimal:

$$x_{11} = 0 \qquad x_{12} = 0 \qquad x_{13} = 80$$

$$x_{21} = 20 \qquad x_{22} = 70 \qquad x_{23} = 30 \qquad \text{Total ton mileage} = 2450.$$

Goldman's Problem: Combining Optimal Locations, Production Levels, and Transportation Flows

The application of linear programming in the last section provided an indication of how decisions on transportation flows, and on the location of land use generating these flows, can be treated simultaneously. The following example demonstrates the remarkable versatility of linear programming as a tool in complex decision making, in this case the

TABLE 4.1 Redistribution of Flows in a Flow Matrix While Maintaining Row and Column Sums

Column →	1	j	k	m	row sums
Row 1	X_{11}	X_{1j}	X_{1k}	X_{1m}	P_1
Row i	X_{i1}	$X_{ij} + f$	$X_{ik} - f$	X_{im}	P_i
Row u	X_{u1}	$X_{uj} - f$	$X_{uk} + f$	X_{um}	P_u
Row n	X_{n1}	X_{nj}	X_{nk}	X_{nm}	P_n
Column Sums →	C_1	C_j	C_k	C_m	

simultaneous determination of origin and destination of shipments, of number and type of shipments, and of production locations and production figures so as to minimize overall transportation costs while meeting a variety of constraints.

Consider three islands which we shall label 1, 2, 3. We will make the following assumptions:

(1) Each island produces a single raw material: iron ore on island 1, coal on 2, and limestone on 3. There is no upper limit on the production of these three raw materials.
(2) The demand for steel on the island is D_i (i = 1, 2, 3).
(3) The raw materials required to produce one unit of steel consist of 1.5 units of iron ore, 4 units of coal, and 0.5 units of limestone.
(4) Cost for each shipment between any two islands is 100 monetary units, irrespective of whether the cargo is some raw material, steel, or an empty shipment.
(5) The cost per unit of steel production is the same on all three islands.

Since our intent is the construction and examination of a model of some real-world segment rather than the study of a real-world situation itself, no parameters other than those listed will be considered here. To simplify notation, we will measure all steel and raw material quantities in shiploads (rather than tons). We wish to find a solution for steel production figures and shipments that minimizes transportation costs C while meeting all demands. Specifically, we pose the following questions:

(a) On what islands should steel be produced, and how much?
(b) How much raw material should be shipped, from where, and where to?
(c) What quantities of steel should be shipped, and what are the origins and destinations of these steel shipments?
(d) How many empty shipments are needed to assure that the transportation activities of questions b and c do not lead to accumulation of ships at particular islands? From where to where should these empty ships be moved?

We will formulate the problem as a linear program. Once this has been accomplished, the solution can readily be computed by one of the existing linear program computer routines.

Starting with question a, let x_i refer to the (as yet unknown) amount of steel to be produced at the island i; i = 1, 2, 3. Notice that the answer to question a in combination with assumptions 1 and 3 provides the answer for question b: For any level of steel production on any one of the three islands, it is clear what raw materials have to be imported, how much of them, and where from.

The answers to questions c and d—how many shiploads of steel and how many empty shipments should be moved between the islands—are, of course, not known at this point, and we will have to treat the numbers of these shipments as variables, with x_{ij} representing the number of steel shipments and y_{ij} the number of empty shipments from i to j, where $i, j \in \{1, 2, 3\}$. With these definitions, we are now prepared to set up the linear program. There are two types of constraints:

(1) Steel demand constraints, one for each island. The local steel production at island i minus exports plus imports to and from the other two islands constitutes the net amount of steel locally available and has to be at least as much as the local demand. For the first island, this condition translates into $x_1 - x_{12} - x_{13} + x_{21} + x_{31} \geq D_1$, or, in general terms for any island i:

$$x_i - x_{ij} - x_{ik} + x_{ji} + x_{ki} \geqslant D_i, \quad \text{where} \quad i \neq j \neq k \neq i; \quad \text{and} \quad i, j, k \in 1, 2, 3 . \qquad [29]$$

(2) Zero sum ship movements: During each time unit, there must be an equal number of ship arrivals and departures at the island i (i = 1, 2, 3)—including, if necessary, some empty shipments—to avoid a

growing number of idle ships accumulating over time. There are three types of shipments: those carrying steel, those carrying raw materials, and the empty shipments.

Take, as an example, island 2. Its steel production is x_2, and it has to import 1.5 shiploads of iron ore from island 1 and 0.5 shiploads of limestone from island 3 for each shipload of steel it produces. Thus, the total number of incoming raw material shipments is $1.5x_2 + 0.5x_2$. Similarly, the island exports coal to the islands 1 and 3, and since their steel productions are, respectively, x_1 and x_3, the number of coal shipments leaving island 2 is $4x_1 + 4x_3$. If we now add the incoming and outgoing steel shipments and empty shipments we can balance the total flow by setting total incoming and total outgoing flows as equal:

$$\begin{aligned} &1.5x_2 + 0.5x_2 + x_{12} + x_{32} + y_{12} + y_{32} \\ &= 4x_1 + 4x_3 + x_{21} + x_{23} + y_{21} + y_{23}. \end{aligned} \qquad [30]$$

Equivalent equations balancing incoming and outgoing ship movements can easily be formulated for the other two islands.

Typically there are many different sets of numerical values for the fifteen variables each satisfying the foregoing constraints. We wish to identify the particular set of production and flow figures that minimizes the total transportation cost involved. Since we have already identified all three types of shipments, it is now quite simple to formulate the objective function:

$$\begin{aligned} \text{Min } C = 100 \{ &4.5x_1 + 2.0x_2 + 5.5x_3 \\ &+ x_{12} + x_{13} + x_{21} + x_{23} + x_{31} + x_{32} \\ &+ y_{12} + y_{13} + y_{21} + y_{23} + y_{31} + y_{32} \} \end{aligned} \qquad [31]$$

This objective function, together with the set of constraints, completely describes our problem as a linear program ready to be solved by any number of available solution methods.

It is not difficult to predict the principal orientation of the outcome: The minimum cost solution will favor a relatively high production figure of steel on island 2 because steel production on islands 1 and 3 requires transportation of coal; coal, however, because of its large input of four units for each unit of steel, requires relatively more, and therefore more expensive, transportation. Also, the minimum cost solution will keep

TABLE 4.2 Constraints of Goldman's Problem in Matrix Form

Variables	x_1	x_2	x_3	x_{12}	x_{13}	x_{21}	x_{23}	x_{31}	x_{32}	y_{12}	y_{13}	y_{21}	y_{23}	y_{31}	y_{32}	
1	+1			−1	−1	+1		+1								$\geqslant D_1$
islands 2		+1		+1		−1	−1		+1							$\geqslant D_2$
3			+1		+1		+1	−1	−1							$\geqslant D_3$
1	+4.5	−1.5	−1.5	−1	−1	+1		+1		−1	−1	+1		+1		= 0
islands 2	−4.0	+2.0	−4.0	+1		−1	−1		+1	+1		−1	−1		+1	= 0
3	−0.5	−0.5	+5.5		+1		+1	−1	−1		+1		+1	−1	−1	= 0
	Steel productions and raw material shipments			Steel shipments						Empty shipments						

empty shipments at a low level or avoid them altogether through a shipping schedule emphasizing exchange of raw and finished products.

For illustrative purposes, Figure 4.2 summarizes a minimum cost example given by Goldman (1958). One can easily verify that the numbers of incoming and outgoing ships are equal for each island port, and that all steel demands are met by local production and, where necessary, by imports of steel from other islands. Note that island 3, with the largest demand for steel by far, does not produce any steel in this minimum cost solution, and that of the total of 6000 shipments, only 200, or 3.3 percent, are empty.

The Simplex Method

The simplex method is a systematic stepwise procedure, that is, an algorithm which, after a finite number of steps, produces an optimal solution for a linear program (Dantzig 1963). It is based on the algebra of systems of linear equations, and to this end we first subject the linear program to several transformations. As we discussed in the first section of this chapter, a linear program consists of a linear objective function and a set of constraints:

$$\text{Max } z = \sum_{j=1}^{m} u_j x_j$$

s.t.: [32]

$$\sum_{j=1}^{m} a_{ij} x_j \geqslant b_i$$

$$i = 1, 2, \ldots r.$$

To prepare the linear program for the application of the simplex method, the following transformations are required:

(1) We convert each constraint imposing a lower limit on a linear combination of the variables into a constraint that defines an upper limit. This can easily be accomplished because an inequality will reverse its sign (from $\leq$ to $\geq$ or reverse) if we multiply it by (-1):

$$\sum_j a_{ij} x_j \geqslant b_i \rightarrow \sum_j (-a_{ij}) x_j \leqslant (-b_i) \qquad [33]$$

or $$\sum_j A_{ij} x_j \leqslant B_i \qquad [34]$$

where $A_{ij} = (-a_{ij})$ and $B_i = (-b_i)$ for all i,j.

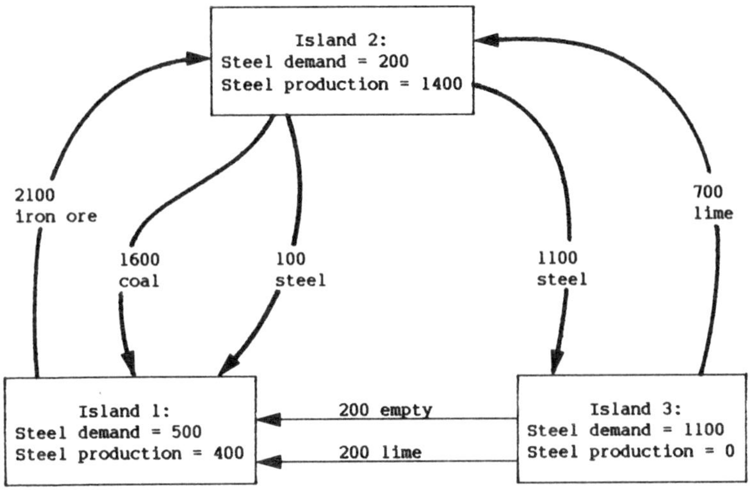

Figure 4.2 Example of a Goldman Problem and Its Numerical Solution

Note: Linear programs which, at the end of this step, have negative constants $B_i < 0$ for some values of i will not be treated here, as their solution involves several additional steps.

(2) Next, we convert the linear inequality (34) into an equality by adding a new variable, y_i, called a slack variable, for all inequalities $i = 1, 2, \ldots, r$:

$$\sum_j A_{ij} x_j + y_i = B_i, \qquad i = 1, 2, \ldots, r. \qquad [35]$$

Since, at this point, all constraints define upper limits, the slack variables are necessarily nonnegative.

(3) We will assume that all variables x_j are larger than or equal to zero, that is, that none of the variables can assume negative values. If a particular variable represents, say, the number of motor vehicles in operation, then our assumption will always be satisfied, as there are no such things as negative vehicles. However, if the problem at hand requires that we allow for negative values of some variable x_j—say, the balance of a financial account—then we represent that variable as the difference of two new variables:

$$x_j = x_j' - x_j'' \quad \text{where} \quad x_j' \geqslant 0,\ x_j'' \geqslant 0$$

The substitution of a variable covering the full range of positive and negative values by two variables each of which will assume only nonnegative values is permissible because any number can always be expressed as the difference of two nonnegative numbers.

(4) Furthermore, we convert each maximization problem into one of minimization by taking the negative of the objective function. For illustration, imagine a straight line segment with a slope $\neq 0$ in two-dimensional space and consider the particular point on it in which it assumes its minimum. If we now take the negative of that line segment—that is, its image when mirrored on the x-axis—that lowest point becomes its highest point, that is, its maximum. Thus, in our case,

$$\text{Max } z = \sum_j u_j x_j \quad \text{becomes} \quad \text{Min } (-z) = \sum_j (-u_j) x_j$$

or, for $(-z) = Z$ and $(-u_j) = U_j$, it translates into $\text{Min } Z = \sum_j U_j x_j$.

The first and fourth transformations may change the signs of coefficients a_{ij} and u_j and of the constants b_i; the second and third transformations will increase the set of variables by r slack variables, and may increase it by additional variables to ensure that all variables are limited to nonnegative values. We will assume the total number of variables to be $r + n$, not counting z, the dependent variable of the objective function.

To reflect the changes, we denote the variables by X_j, $j = 1, \ldots, r + n$, and define the slack variables as $y_i = X_i$, $i = 1, 2, \ldots, r$. The coefficient of the variable X_j in the ith constraint will be labeled A_{ij}, and will be labeled U_j in the objective function. The dependent variable of the objective function will be represented by Z, and the constraining constants, originally denoted by b_i, will now be labeled B_i. Summarizing, our linear program (32), after execution of the transformations, takes on the form

$$\text{Min } Z = \sum_{j=1}^{r+n} U_j X_j \qquad (\text{with } U_j = 0 \text{ for } j = 1, \ldots, r)$$

$$\text{s.t.:} \qquad [36]$$

$$\sum_{j=1}^{r+n} A_{ij} X_j = B_i, \qquad i = 1, 2, \ldots, r.$$

By construction, all variables X_j are limited to nonnegative values, and by assumption, all constants B_i are also nonnegative. Since the coefficients of the slack variables are all equal to 1, we can rewrite the linear program in the following format:

$$\begin{array}{llllll} X_1 & & & + A_{1,r+1} X_{r+1} + \cdots + A_{1,r+j} X_{r+j} + \cdots + A_{1,r+n} X_{r+n} = B_1 \\ & X_2 & & + A_{2,r+1} X_{r+1} + \cdots + A_{2,r+j} X_{r+j} + \cdots + A_{2,r+n} X_{r+n} = B_2 \\ & & X_r & + A_{r,r+1} X_{r+1} + \cdots + A_{r,r+j} X_{r+j} + \cdots + A_{r,r+n} X_{r+n} = B_r \\ \hline & & & -Z + U_{r+1} X_{r+1} + \cdots + U_{r+j} X_{r+j} + \cdots + U_{r+n} X_{r+n} = 0 \end{array} \qquad [37]$$

Including the objective function (last row), the system consists of r + 1 equations in r + n + 1 variables. Note that for $i < r + 1$ the coefficient of the ith variable is one in the ith equation and zero in all others. Such variables are called basic, while all others are called nonbasic; a system of linear equations represented in this form is called a canonical system. Its particular form permits us to read off immediately a first feasible solution known as the "basic" solution of the canonical form: Assign to all basic variables X_i the value B_i where $i = 1, \ldots, r$; assign to all nonbasic variables X_j the value 0 where $j = r + 1, \ldots, r + n$, and set $Z = 0$.

Moreover, if $U_{r+j} \geq 0$ for all j, then this feasible solution is also minimal because, by construction, the associated variables X_{r+j} cannot have negative values, and assigning to any one of them a positive value would increase the value of Z, making it suboptimal. Thus, our first feasible solution can be improved only if at least one of the coefficients U_{r+j} in the objective function is negative. Let us therefore assume that one or more coefficients in the objective function are negative, and let U_{r+k} be the smallest one of them:

$$U_{r+k} = \text{Min}\,\{U_{r+j} \mid U_{r+j} < 0;\ j = 1, 2, \ldots, n\}. \qquad [38]$$

We lower the value of Z by assigning some positive value to X_{r+k}, say X'_{r+k}, while keeping the other values of our first feasible solution unchanged. The new solution produces a lower Z value but is no longer feasible, since the constraining equations are now unbalanced. Take, for example, the ith equation:

$$X_i + A_{i,r+1}X_{r+1} + \ldots A_{i,r+k}X_{r+k} + \ldots + A_{i,r+n}X_n = B_i. \qquad [39]$$

In our first solution we had $X_i = B_i$, with all other variables in that equation equal to zero. Assume that $A_{i,r+k} > 0$. If we give X_{r+k} a positive value, we can maintain the equation only by lowering one of the other variables. Since none of them can be less than zero, only the basic variable X_i can be lowered. The lowest permissible value of X_i is zero, in which case $A_{i,r+k}X_{r+k} = B_i$. Hence, $X_{r+k} = B_i / A_{i,r+k}$ is the largest amount we can choose for X_{r+k}, at least inasmuch as the balancing of equation i is concerned. Likewise, other equations may limit the size of the numerical value we can assign to X_{r+k}, and its maximum possible value is therefore the minimum limit imposed by the constraining equations:

$$\text{Max } X_{r+k} = X'_{r+k} = \text{Min}\left\{B_i/A_{i,r+k} \mid A_{i,r+k} > 0;\ \ i = 1, \ldots, r\right\} \qquad [40]$$
$$= B_t/A_{t,r+k}.$$

However, in the special case that none of the coefficients $A_{i,r+k}$ $(i = 1, \ldots, r)$ is positive, we can choose any positive value for X_{r+k} and adjust the values of the basic variables X_i $(i = 1, \ldots, r)$ to balance the equations. In this case there is no lower limit for the value of Z.

Notice that we cannot increase the value of X_{r+k} or that of any other nonbasic variable if at least one of the constants B_i is zero. In this case we call the basic solution degenerate. Let us therefore assume that $B_i > 0$, and that at least one of the coefficients $A_{i,r+k} > 0$, for $i = 1, \ldots, r$.

Assigning X_{r+k} its maximum possible value X'_{r+k} and lowering the values of the basic variables accordingly will give us a second, improved feasible solution:

$$X_i = B_i - A_{i,r+k} X'_{r+k} \qquad \text{for } \ i = 1, \ldots, r; \quad X_t = 0$$
$$X_j = 0 \qquad \text{for } \ j = r+1, \ldots, r+n \ \text{ and } \ j \neq r+k \qquad [41]$$
$$X_{r+k} = X'_{r+k} = B_t/A_{t,r+k}; \quad Z = U_{r+k} X'_{r+k} < 0.$$

We derived the first feasible solution—the basic solution of canonical system 37—by setting each basic variable equal to the (positive) constant of the corresponding equation and setting all nonbasic variables equal to zero. In our improved solution, however, one of the basic variables, X_t, is zero, and one of the nonbasic variables, X_{r+k}, is nonzero. We will therefore rearrange our system of equations 37 so that (I) it is again in canonical form, and (II) our improved solution is the basic solution of this form.

We do so by making X_{r+k} a basic variable with unit coefficients in row t and zero coefficients elsewhere. To this end, we divide row t by $A_{t,r+k}$ and, subsequently, multiply it by $A_{i,r+k}$ and subtract it from row i, where $i = 1, \ldots, r + 1$ and $i \neq t$. Next we exchange columns t and r + k and change the subscripts of the coefficients and variables in them so that they conform to the standard labeling (namely, starting in the upper left corner, the element in the ith row and j^{th} column carries the subscripts i,j). These last two steps are not essential, but for convenience only.

The system is now again in canonical form, and by construction, its basic feasible solution is identical to our second, improved solution. If, at this point, none of the coefficients of the objective function is negative, no further improvement of the solution is possible; otherwise we return to equation 38 and repeat the process of improving the last basic feasible solution. We continue with the iterative process until we either encounter a situation of degeneracy, as described before, or else all coefficients of the objective function are nonnegative, indicating optimality of the last improved solution.

Example: Let us solve the transportation problem presented in the first section of this chapter by means of the simplex method. However, to keep the number of slack variables down, we will include only three of the constraints. In particular, so as to ensure that we end up with the same result, we include the two constraints 1 and 3b in the linear program of equation 22. In addition, we include the fourth constraint. Thus, together with the objective function, our linear programming problem can now be stated as

$$\text{Max } P = 700x + 1400y$$

$$\text{s.t.:} \qquad [42]$$

$$x + y \leqslant 9; \quad y \leqslant 5.5; \quad 7x + 3y \leqslant 49$$

Note that the inequalities establish nonnegative upper limits as required by the simplex method. Moreover, we can restrict ourselves to nonnegative values for x and y, as negative flow volumes would be meaningless. However, we need to transform the objective function into a minimization problem: Min Z = (–P) = –700x – 1400y. Next we set

$x = x_1$, $y = x_2$, and add three slack variables to change the inequalities into equations:

$$\begin{aligned} x_3 \quad + \quad x_1 + \quad x_2 &= 9 \\ x_4 \quad + \quad\quad + \quad x_2 &= 5.5 \\ x_5 \quad + \quad 7x_1 + \quad 3x_2 &= 49 \\ -Z - 700x_1 - 1400x_2 &= 0. \end{aligned} \qquad [43]$$

At this point the problem is stated as a canonical system (equations 37) and is ready for the application of the simplex method.

I. A first feasible solution is the so-called basic solution of canonical system 43; it is defined by setting the basic variables and Z equal to the equation constants, and setting the nonbasic variables equal to zero:

$$x_3 = 9; \quad x_4 = 5.5; \quad x_5 = 49; \quad x_1 = x_2 = Z = 0. \qquad [44]$$

II. We improve the first feasible solution by selecting the smallest negative coefficient in the objective function—that is, –1400—and assigning to the corresponding variable x_2 the largest positive value permitted by the constraints.

We keep in mind that in the first feasible solution $x_1 = 0$, and that only nonnegative values of the variables are permitted. It follows that the maximum permissible value of x_2 in the first constraint is 9, in which case we have to change the value of x_3 to zero. Similarly, the maximum possible value of x_2 is 5.5 in the second constraint and 49/3 in the third constraint, with $x_4 = x_5 = 0$. To guarantee that no variable will be assigned a negative value, we set $x_2 = 5.5$, that is, the minimum of the permissible values 9, 5.5 and 49/3. Recalculating the values of the other variables in canonical system 43 produces the first improved feasible solution:

$$\begin{aligned} &x_1 = 0; \quad x_2 = 5.5; \quad x_3 = 3.5 \\ &x_4 = 0; \quad x_5 = 32.5; \quad Z = -7700. \end{aligned} \qquad [45]$$

Notice that the improved solution 45 is not the basic solution of canonical system 43. Now one of the basic variables, x_4, is zero, and one of the nonbasic variables, x_2, is positive. To exchange x_4 and x_2 as basic and nonbasic variables, we multiply the second equation in canonical system 43 by –1, by –3, and by 1400 and add it, respectively, to the first, third, and fourth equations of 43. We then exchange the x_4– with the x_2–column and get the following result:

$$\begin{aligned} x_3 \quad + \quad x_1 - \quad x_4 &= 3.5 \\ x_2 \qquad\qquad + \quad x_4 &= 5.5 \\ x_5 \quad + \quad 7x_1 - \quad 3x_4 &= 32.5 \\ -Z - 700x_1 + 1400x_4 &= 7700. \end{aligned} \qquad [46]$$

Our system of equations is now again stated in the canonical format; moreover, its basic solution—setting the basic variables and Z equal to the equation constants and setting all nonbasic variables equal to zero—is now identical to our first improved solution 45. We are therefore ready for the next round of improving our solution by repeating step II.

The smallest negative coefficient in the objective function of canonical system 46 is –700. We will give the corresponding variable x_1 the largest permissible positive value. In the first equation of canonical system 46, x_1 can be as large as 3.5 by reducing the value of x_3 to zero while keeping $x_4 = 0$. In the second equation, x_1 can assume any positive value because its coefficient is zero. In the third equation, x_1 can be at most 32.5/7, in which case $x_5 = x_4 = 0$. Hence, the maximum possible value of x_1 is 3.5. Substituting this value in canonical system 46 and keeping in mind that x_4 remains zero, we get the second improved solution:

$$\begin{aligned} &x_1 = 3.5; \quad x_2 = 5.5; \quad x_3 = 0 \\ &x_4 = 0; \quad x_5 = 8; \quad Z = -10{,}150. \end{aligned} \qquad [47]$$

To improve further our second improved solution 47, we transform canonical system 46 such that x_3 becomes a nonbasic variable and x_1 changes into a basic variable. This we accomplish by multiplying the first equation in the system 46 by –7 and +700 and adding it, respectively, to the third and fourth equations. Subsequently exchange the x_3 and x_1 columns, with the following result:

$$\begin{aligned} x_1 \quad + \quad x_3 - \quad x_4 &= 3.5 \\ x_2 \qquad\qquad - \quad x_4 &= 5.5 \\ x_5 \quad - \quad 7x_3 + \quad 4x_4 &= 8 \\ -Z + 700x_3 + 700x_4 &= 10{,}150. \end{aligned} \qquad [48]$$

By construction, the basic solution of canonical system 48 is equal to our second improved solution 47. At this point the coefficient in the objective function are all positive. To reduce the value of Z even further, we would have to change the values of x_3 or x_4 or both. Since their values in the last solution 47 are zero, and since only nonnegative values are permitted, solution 47 cannot be improved, and $x_1 = 3.5$, $x_2 = 5.5$, and $P = -Z = \$10{,}150$ is therefore the optimal solution.

5. DESIGN AND OPERATION OF NETWORKS

Routes and Networks

In this chapter we explore the relationship between the spatial pattern of transportation demand and the spatial layout and utilization of the transportation network that is supposed to meet this demand.

In a country with a nice climate and little air traffic, each plane will simply fly straight to its destination. Routing is therefore a simple matter: To save time and cost, each flows follows a straight line. That is usually no the case for ground transportation, for several reasons:

(1) The ease of transportation flows—namely, the energy required to overcome distance—varies with terrain and the physical condition of the surface on which transportation movement takes place.
(2) Existing land use of a particular area may be incompatible with the use for transportation flows—in fact, this applies to almost all land uses other than transportation itself. Hence, land areas in the form of corridors have to be set aside to accommodate transportation flows.
(3) Land is a finite resource and, at least in urban areas, in high demand, and any plan to provide each pair of origins and destinations with a nearly straight transportation route would consume an inordinate amount of land at the expense of other land uses needed by society.

As we shall see, it is the outstanding accomplishment of networks that they provide reasonably good transport connections between large numbers of diverse origins/destinations with comparatively few links and a relatively small overall network length.

Moving people and freight consumes energy, and a large part of transportation research deals with the question of how the demand for transportation can be met with a minimal energy expenditure. One obvious answer, which even nature applies extensively, is the separation of the transportation effort into two components: to spend some energy on the preparation and maintenance of the route of movement, and whatever additional energy is required for the movement itself. Note that the two energy figures are inversely related—at one extreme we have cross-country transportation with minimal, if any, route preparation but high cost of movement per ton mile (for example, four-wheel trucks moving through wilderness terrain), and at the other extreme high cost of route preparation and very low cost of movement (for example, water canals with transportation by barges). Every form of ground transportation is positioned somewhere between these two extremes.

In an urban area, roads, railways, and especially subways consume enormous amounts of resources for construction and maintenance but permit the cheap and fast transportation of millions of people or tons of freight. A simple quantitative consideration will underscore our argument.

Let f be the volume of annual flow between two locations measured, say, by number of vehicles; let k be the cost of movement per vehicle mile; and let c be the cost of route preparation and maintenance per mile, on an annual basis. Then the combined annual cost of route preparation and transportation flow per mile is $c + f \cdot k$. While the flow cost $f \cdot k$ is usually borne by the user, the cost of route preparation is paid by governmental agencies that obtain their funds through taxes (for example, gasoline tax). That is, at least in our simplified approach, the user pays the entire bill—either out-of-pocket or in the form of taxes. If each user operates one vehicle, then the combined cost C of preparation and flow per mile per user is

$$C = \frac{c}{f} + k; \qquad [49]$$

that is to say, the more users the lower the cost per user. Just like mass production in industry, mass movement in transportation will lower the cost per unit, in this case, per mile and user.

However, given the spatial dispersion of people and their activities in an urban area, how can we generate mass movements between different locations? That can be accomplished by building a system of feeder routes that will collect individual flows and bundle them into mass flows carried by trunk routes. Another system of feeder lines will eventually permit the disaggregation of the mass flow to enable each user to reach his or her particular destination. We call the strategy of bundling individual flows into mass flows by means of feeder and trunk lines the network principle.

To be sure, there are limits to this strategy: Building feeder lines and moving individual users with different origins and destinations together along the same route for at least part of their respective trips produces roundabout routes for most users, that is, more or less substantial deviations from a straight-line movement to their destinations. The stepwise aggregation of individual flows into large mass flows means increased roundabout ways for many individual flows and therefore increased flow cost, which reduces the savings generated by the mass movement as explained above. Thus, the planner is confronted with the task to find an "optimal" solution by determining the level of flow aggregation that will minimize the joint cost of network construction and maintenance, on one side, and the movement of flows, on the other.

The following sketch will demonstrate that the principle of bundling individual flows on a system of interconnected routes is exceedingly cost-effective and one of the great organizing schemes both in nature and in society.

The Network Principle: An Example

Consider a 10×10 square grid network, the side of each square being one mile. The network consists of 100 nodes and 180 links and has an overall length of 180 miles. For the sake of simplicity, let us assume that the amount of traffic between all pairs of nodes is the same. Those pairs of nodes that are located in positions parallel to one of the grid axes will be provided with a straight-line network connection; flows between all other pairs of nodes must deviate from the straight line. At worst, origin and destination of a flow are positioned on a line diagonal to the grid system—for example, the nodes A and B in Figure 5.1.

We know from elementary geometry that if the straight distance between A and B is (AB), then the distance [AB] over the grid network—the so-called city block distance—is (AB) times the square root of 2; $[AB] = (AB)\sqrt{2} = 1.41(AB)$ and is therefore about 41 percent longer than the straight-line distance. On average, for any pair of nodes A, B, the increase is substantially less but still in the order of 27 percent.

To avoid this circuitous routing, we replace the grid network with a complete network providing each pair of nodes with a straight-line road connection. There is a total of $\frac{1}{2} \times 100 \times 99 = 4950$ different pairs of nodes, the average (straight) distance being approximately $5\frac{1}{2}$ miles, giving the network a total length of 27,225 miles. To summarize:

	Network Mileage	Ratio of Network Distance to Straight Distance
Square Grid Network	180	1.27
Complete Network	27225	1.00
Ratio	0.006	1.27

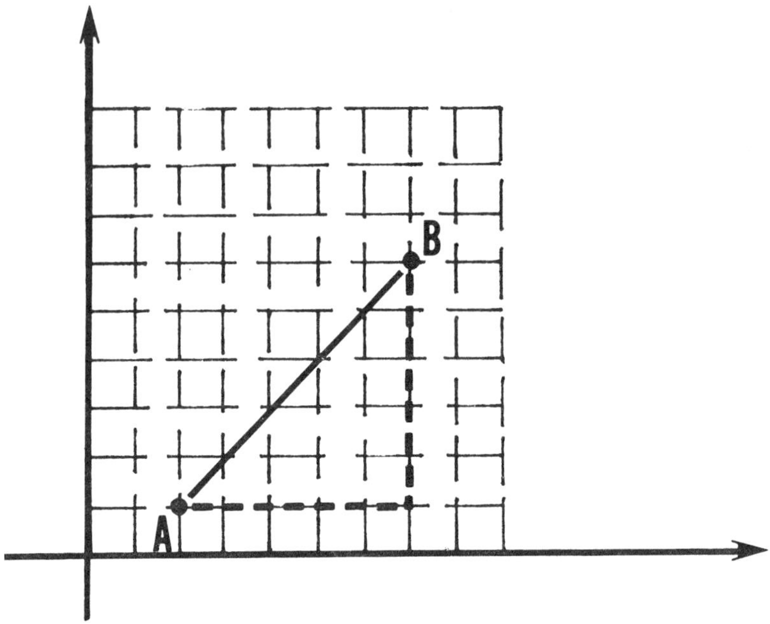

Figure 5.1 Straight Line Versus City Block Distance in a Square Grid Network

In words, on 180 miles of road the grid network provides road connection between all pairs of nodes, each connection being, on average, 27 percent longer than a straight-line connection would be, and it provides this level of service with a network mileage that is far below 1 percent of the length of the complete network consisting of straight-line connections for all pairs of nodes. Thus, while the flow mileages of the two networks differ only by a factor of 1.27, the respective road mileages differ by a factor of 150.

Notice that the flows of the grid network exhibit two features characteristic of network flows in general: A single flow typically uses several different links, and many different flows use the same link in getting to their respective destinations. Neither of these two statements applies to the completely connected network: Here the individual flows correspond to the individual links and vice versa; links are independent of one another, as their interconnection by nodes is strictly formal but not functional. For large numbers of network nodes, the length of complete networks is astronomical and, in reality, unworkable. In fact, most real road systems, both natural and man-made, could not have developed without extensive application of the network principle discussed in the previous section.

Optimal Route Design I: The Law of Refraction

Transportation networks are composed of links interconnected in nodes. Before we address the overall design of a network, we will first present two models that permit the optimal design of individual links.

Most models generating optimal routes for future road or rail construction identify a sequence of points—usually nodes of existing networks—to be connected by essentially straight transportation links. However, straight routes are frequently suboptimal, as is particularly obvious when mountainous terrain has to be crossed or expensive land acquisition can be avoided. Of course, deviation from a straight route will increase two other cost figures directly related to the overall transportation costs: It will increase the number of miles that need to be built, and it will increase the number of vehicle miles of the traffic using the new route. The following model provides an example of how a transportation line can be routed through an area of differential construction cost such that the combined cost of construction and flow is minimized.

For maximum simplicity, we will assume that the area A to be traversed can be divided into two subareas, A_1 A_2, with associated annual construction and maintenance costs c_1, c_2 per mile of roadway, where $c_1 > c_2$. We further assume that the planned road has to connect two fixed locations, $P_1 \epsilon A_1$ and $P_2 \epsilon A_2$, and will carry an annual traffic volume f. Finally, we approximate the boundary between A_1 and A_2 by a straight line. (Any one of these assumptions can be relaxed so as to broaden the applicability of the model; see Werner 1968.)

We superimpose a system of Cartesian coordinates on this geographic setting and place it, for reasons of mathematical convenience, so that the x-axis coincides with the boundary between A_1 and A_2 and P_2 is a point located on the y-axis (Figure 5.2).

Since the subareas A_1 and A_2 are supposed to be internally homogeneous with regard to road construction and maintenance cost per mile, it

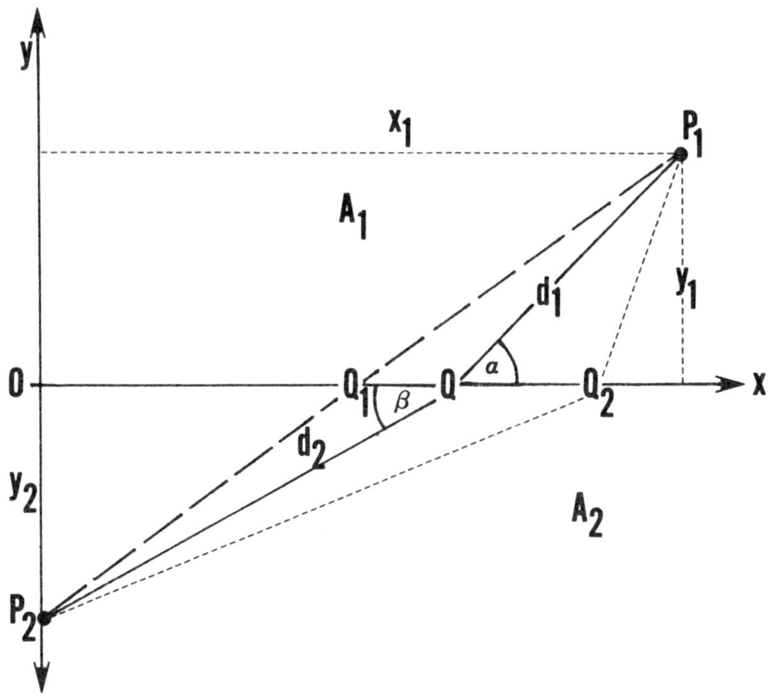

Figure 5.2 Alternative Transportation Routes Crossing Terrain of Differential Construction Costs

is clear that, to minimize overall costs, the two segments of the future route connecting P_1 and P_2 should be straight in these two subareas. The question is: Should the entire route be straight (Figure 5.2, route $P_1Q_1P_2$), or, if not, at what point Q should it cross the boundary between A_1 and A_2?

It seems plausible to select a route that has a relatively short segment in the more expensive area A_1 and therefore, by necessity, a relatively longer segment in A_2 (Figure 5.2, route $P_1Q_2P_2$). On the other hand, moving the critical crossing point from Q_1 to Q_2 will increase the overall mileage that needs to be constructed and, of course, also the mileage for the traffic flow. The problem we are faced with, therefore, is to determine the exact point Q at which the overall transportation cost assumes its minimum value.

As is readily apparent from Figure 5.2, the total flow cost C_f is equal to the flow volume f times the cost per unit flow per mile, k, times the length of the route, $(d_1 + d_2)$: $C_f = k \cdot f(d_1 + d_2)$. The total construction cost C_c consists of the construction cost figures in the two subareas: $C_c = c_1d_1 + c_2d_2$. According to the Pythagorean theorem, it is

$$d_1 = \sqrt{(x_1 - x)^2 + y_1^2}; \quad d_2 = \sqrt{x^2 + y_2^2} \qquad [50]$$

where (x_1, y_1) and $(0, y_2)$ are the coordinates of the two locations P_1 and P_2, and $(x, 0)$ are the coordinates of the point Q at which the route will cross the x-axis representing the boundary between A_1 and A_2. Thus, the total cost C_T is

$$\begin{aligned} C_T &= C_f + C_c \\ &= (c_1 + fk)\sqrt{(x_1 - x)^2 + y_1^2} + (c_2 + fk)\sqrt{x^2 + y_2^2} \end{aligned} \qquad [51]$$

and our problem can now be restated in mathematical terms: Find the value of x for which the total cost C_T is minimized. We will solve the problem first graphically and then with the help of calculus.

To generate a graphical display of the cost function C_T let us study it in more detail. Clearly, the expressions in parentheses are simply constant coefficients—they represent the total cost of construction and flow per mile in each of the two areas A_1 and A_2. Since the square roots represent the respective lengths of the two links of the route from P_1 to P_2, equation 51 expresses the total cost, C_T, as the sum of the total cost of each link: $C_T = C_1 + C_2$. Notice that C_1, the cost for the first link, will assume its minimum for $x = x_1$ and will grow monotonically as x becomes larger or smaller than x_1 (since the roots represent road length, we will consider only their positive values). Similarly, C_2 assumes its minimum for $x = x_2 = 0$ and will be the larger the more x deviates from 0.

In Figure 5.3 we have plotted the curves of the two cost functions C_1, C_2. Adding these two functions graphically by adding their respective values for each value of x gives us the curve of the function C_T. We can now read off the minimum cost solution directly from our graph: The overall cost function C_T assumes its minimum value for $x = x_Q$.

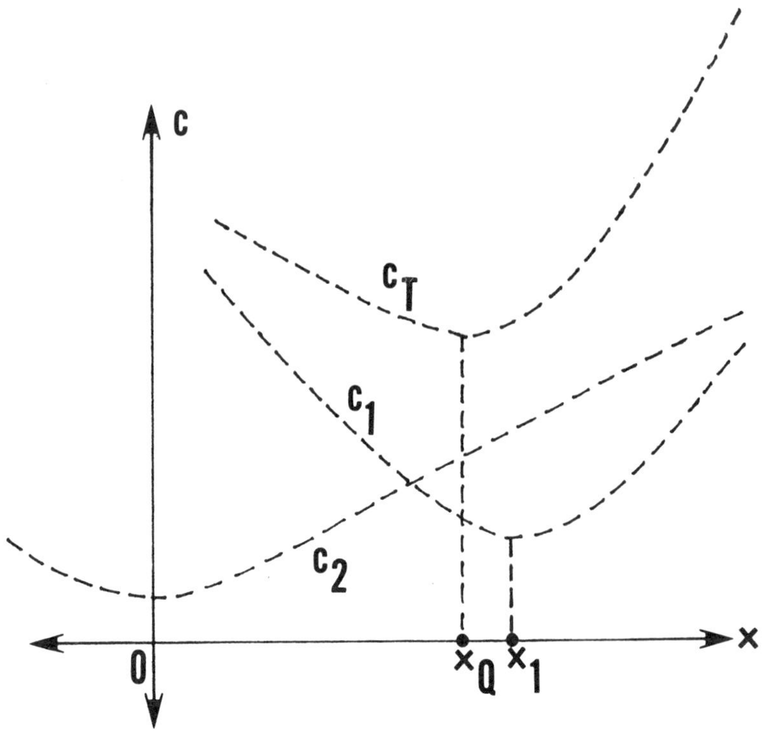

Figure 5.3 Joint Cost of Construction and Traffic Flow as a Function of Route Location as Shown in Figure 5.2

Alternatively, we can solve our problem through differential calculus. To this end we determine the first derivative of the cost function C_T with regard to x:

$$\frac{dC_T}{dx} = -(c_1 + fk) \frac{(x_1 - x)}{\sqrt{(x_1 - x)^2 + y_1^2}} + (c_2 + fk) \frac{x}{\sqrt{x^2 + y_2^2}}. \quad [52]$$

Note that in Figure 5.2 the first ratio on the right-hand side of equation 52 represents the cosine of the angle α at which the route coming from P_1

meets the x-axis, and the second ratio represents the angle β which the second segment of our route forms with the x-axis. Furthermore, the associated coefficients $(c_1 + fk)$ and $(c_2 + fk)$ constitute the total unit cost in the two subareas, that is, the combined costs of both construction and flow per mile and year.

The cost function will assume its minimum for that particular value of x for which the first derivative of the cost function C_T is equal to 0:

$$-(c_1 + fk) \cos \alpha + (c_2 + fk) \cos \beta = 0 \quad \text{or} \quad \frac{\cos \alpha}{\cos \beta} = \frac{c_2 + fk}{c_1 + fk}. \qquad [53]$$

In words: The total cost will be minimal when the ratio of the total unit costs in the two areas is equal to the ratio of the cosines of the two angles at which the route crosses the boundary between the two areas.

Incidentally, this solution is known as the law of refraction in transportation geography because equation 53 is formally identical to the mathematical equation that describes the refraction of a wave as it moves from one medium into another—for example, a ray of light passing from air into water.

To establish reliably that the point Q determined by this solution does indeed minimize the cost function and that there is no other point on the x-axis in which C_T is minimal, we would have to examine the second derivative of the cost function and prove it to be positive throughout. While this does not pose any problems, we will instead, in the interest of brevity, subject the solution to a partial sensitivity analysis.

If we keep the unit construction cost figures c_1, c_2 and the unit flow cost k constant and increase the flow volume f, then the ratio on the right of equation 53 will move toward 1, and the difference of the cosines of the two angles, and therefore of the angles themselves, will shrink. As a result, the route from P_1 to P_2 will approximate a single straight line. That is to be expected because with increasing flow volume, the flow cost increases relative to the construction cost, and the savings generated by a routing that shortens the road segment in the expensive area A_1 will be more than absorbed by the increased flow cost.

Another example: Let f, k, c_2 be given constants, and let us assume that c_1 will become very large. In this case the ratio $\cos \alpha / \cos \beta$ will approach 0, meaning that $\cos \alpha$ will approach 0 or α will become a 90° angle. That is to say, if construction costs in the subarea A_1 become extraordinarily expensive relative to all other costs, then the first route segment coming from P_1 should reach the boundary at a right angle, that is, in the shortest possible distance. Again, the solution corresponds to our expectation—the difference, of course, being that now we have a result which, under the given assumptions, is both exact and proven, rather than inexact and hypothetical.

Extensions of the model include the treatment of several subareas A_i, each with its own unit road cost, and nonlinear approximations to the boundaries separating them. While the minimum cost route will now consist of a sequence of straight segments, the principle of refraction continues to apply wherever the route crosses a boundary between areas of different unit costs.

Optimal Route Design II

As indicated earlier, virtually all research efforts to determine the "optimal" spatial layout for a new transportation route approach the problem by identifying a finite number of points of that route. These points are then interconnected by straight lines. The resulting route is an open polygon which is fixed in space by the location of its vertices. The following example will serve as a demonstration of the versatility and scope of this method.

Consider a residential area A adjacent to a freeway F (Figure 5.4). A is opened up for traffic by a set of parallel and equidistant residential streets that will be interlinked by a single highway, f(x), connecting U, the point in A farthest away from the freeway, with V, the freeway on/off ramp. We will consider only trips made by car and assume that all traffic moves from residence via residential streets and the new highway to the freeway or the reverse (these assumptions might seem overly restrictive but are actually quite plausible, since they capture most of the traffic in A and virtually all of it during rush hours).

Let c be the annual cost of construction and maintenance per highway mile—briefly referred to as construction cost—and k be the cost per vehicle mile. The amount and spatial distribution of trips generated in A is assumed to be given by a trip generation function Q(x,y) defined over A. We wish to find the particular routing of the highway f(x) that minimizes the joint cost of highway construction and vehicular traffic in A.

To simplify the mathematical formulation and treatment of the problem, we approximate the boundary of A by two integer-valued step functions, g(x), h(x). We superimpose a system of Cartesian coordinates on the area A such that the residential streets are located parallel to the y-axis. For convenience, we also assume that the distance between the streets is one unit of length; thus, each integer value of x corresponds to one particular street, provided that $x_U < x < x_V$, where x_U, y_U and x_V, y_V are the coordinates of the highway end points U and V.

Next we subdivide the area A into cells forming a square grid, with their sides parallel to the axes of the coordinate system. The center

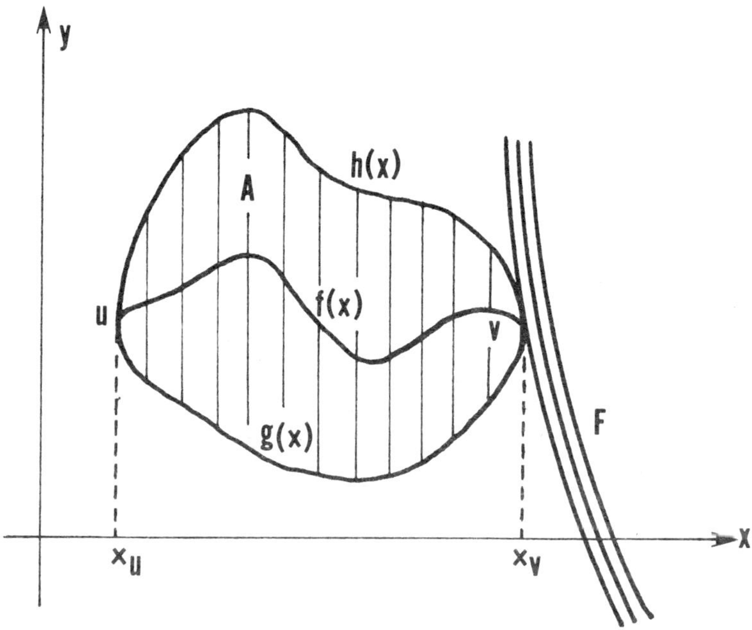

Figure 5.4 Routing of Highway Through Residential Area

points (x,y) of the cells are defined as all and only the points in A whose coordinates are integers. We group the data provided by the density function Q(x,y) in such a way that Q can be redefined as a discrete function that assigns to each center point the number of trips originating or terminating in the corresponding cell, and is of zero value everywhere else. We call the center points centroids and stipulate that all traffic into or out of a cell originates or terminates in its centroid.

The total cost C of construction and flow can be broken down into three components: C_1, the cost of highway construction (including maintenance); C_2, the cost of traffic movement along the residential streets between the individual cells and the highway; and C_3, the cost of the traffic on the highway between U and V. We will determine each cost component separately.

Under the rather simple assumptions of our example, the cost of construction is given by the product of the construction cost per mile times the length of the highway f(x) between U and V. The route of the highway is an open polygon: The intersections with the residential streets x are its inner vertices with coordinates [x, f(x)]; U and V with coordinates (x_U, y_U) and (x_V, y_V) are its outer vertices. Therefore,

$$C_1 = c \cdot \sum_{x=x_U}^{x_V - 1} \sqrt{1 + [f(x+1) - f(x)]^2}. \qquad [54]$$

To calculate C_2, the cost of transportation flow on the residential streets, we multiply the distance from each centroid (x,y) to the highway, $|y - f(x)|$, by the number of trips generated in that centroid, Q(x,y), and determine the sum of these products for all centroids in A. This can be accomplished by first summing $Q(x,y) \cdot |y - f(x)|$ for all integer values of y between g(x) and h(x) for a particular value of x, that is, collecting the individual flows will tend to choose. Loading the flows on the corre- single residential street x. Then, by summing over all streets x between x_U and x_V, we will get the vehicle miles traveled annually on all residential streets, and therefore

$$C_2 = k \sum_{x=x_U}^{x_V} \sum_{y=g(x)}^{h(x)} Q(x,y)\,|y - f(x)|. \qquad [55]$$

In order to obtain the third cost component—the cost of the total vehicle miles traveled on the highway—we multiply the flow volume from each residential street x by the highway mileage as measured between the intersection of this street with the highway and the freeway ramp at V. The flow volume generated by a particular street x is the sum of all values of Q(x,y) between g(x) and h(x), and the distance traveled by this flow along highway f(x) is the length of the polygon f(x) between the vertices [x, f(x)] and V. Thus, the highway mileage associated with trips generated along the street x is

$$\sum_{y=g(x)}^{h(x)} Q(x,y) \cdot \sum_{\xi=x}^{x_V - 1} \sqrt{1 + [f(\xi+1) - f(\xi)]^2}. \qquad [56]$$

We are now ready to determine C_3 explicitly. We sum the highway mileage figures of the flows generated by the various residential streets and do so for all streets between U and V. This sum, multiplied by the cost per unit flow per mile, k, constitutes the total flow cost on the highway:

$$C_3 = k \cdot \sum_{x=x_U}^{x_V} \sum_{y=g(x)}^{h(x)} \sum_{\xi=x}^{x_V} \sqrt{1 + [f(\xi + 1) - f(\xi)]^2}\ Q(x, y). \qquad [57]$$

The problem, minimize $C = C_1 + C_2 + C_3$, is now completely formulated and ready for solution. The route f(x) is completely determined once we find the points at which f intersects with the residential streets such that the cost function C is minimized. The number of unknowns in this problem is therefore equal to the number of residential streets in A.

A crude solution procedure would be to consider several points on each street as possible locations for the intersection with f. Taking one such point from each street will define a route for f, and the cost function C permits the calculation of the expenses associated with it. Other sets of possible intersections lead to different cost figures, and it is easy to see that a computer routine will be able to determine the route of minimum overall cost as long as the number of alternatives is kept manageable.

Network Evaluation and Improvement

Adjusting a transportation network so that it will provide adequate service for the current or expected demand pattern requires reliable estimates about the routes of the various traffic flows on the current network and its future alternatives. These estimates can be generated with the help of the shortest path algorithm presented in the last section of Chapter 3. In this section, we review the application of this algorithm to the problem of optimal network improvement.

Let us consider a transportation network in some urban area serving a particular mode (or modes). Suppose that flow distribution and modal split models have provided us with a complete origin-destination matrix of future flows that need to be accommodated by this network. The shortest path algorithm will give us the particular network routes the individual flows will tend to choose. Loading the flows and the corresponding shortest routes in the network will generate, for each link, the combined volume of all flows using this particular link. Comparing the flow volumes assigned to each link with the link capacity will make it clear where congestion will have to be expected unless the network is modified through appropriate link- or link capacity addition.

There is currently no model available that will tell us where and how much network improvement is required to meet the projected demand pattern in some optimal way. Of course, one might simply increase the capacity of each link up to the level of the expected flow volume; such a solution, however, solves each congestion problem locally and fails to take into account the interdependence of network links. As a result, it might not be a cost-effective solution, in the sense that other solutions might produce a larger decrease of the overall flow cost for each dollar invested in network improvement.

Fortunately, however, the shortest path algorithm permits the evaluation of each network improvement project, as well as any combination of such projects. By way of simulation, the expected flow pattern is loaded on the improved network, and the savings generated by the improvement—whether measured in vehicle miles or man-hours—is then compared with the cost of the improvement project. The same economic evaluation is carried out for alternative improvement projects, and eventually a decision is made as to which project seems to be appropriate within the given budgetary and other constraints.

Similar algorithms have also been used in the design of new networks. In planning the San Francisco Bay Area subway (Bay Area Rapid Transit, or BART), researchers had to determine the location of subway stations in such a way that various competing objectives could be served subject to a given set of legal, financial, and other conditions. For example, minimization of construction cost and maximization of operating speed favor a relatively small number of stations, while maximization of access to the subway favors a large number of subway stations. The shortest path or minimum time algorithm permits the calculation of individual as well as overall travel time of subway users to and from the subway for a variety of different assumptions about both number and location of subway stations.

The evaluation procedure as presented above has a number of shortcomings that need to be addressed. First, the evaluation is restricted to direct costs and benefits, that is, the cost for the network improvement project and the savings generated for the network user. Indirect costs and benefits, such as pollution or change of land values and economic activity, do not enter in the evaluation process; their assessment will be the subject of the last chapter.

Second, the method is critically dependent on the reliability of the route choice predicted by the shortest path algorithm. However, the shortest or fastest route is not always known to the network user, or other considerations may enter the user's route choice. This problem will be addressed at the end of this section.

Third, while a major obstacle to the application of this and similar algorithms—the amount of computing time required to handle larger networks—has been sharply reduced by recent technological advancement, the data requirements can still be formidable. In practical applications, therefore, both the pattern of traffic generating zones and the network interconnecting them are drastically simplified in order to reduce the amount of input data and computing time required. Up to a certain point of generalization, the loss of either accuracy or detail of the model output is at least bearable.

Instead of coding the location of every driveway or parking lot, individual subareas are delineated and computationally represented as single nodes—the centroids—that serve as substitute origins or destinations of all trips starting or terminating in these subareas (for an example, see the previous section). Likewise, the network is stripped of small feeder lines and reduced to a system of major transportation routes interconnecting the centroids. The algorithm will then provide the minimum time paths for this reduced network connecting pairs of centroids.

As is the case with the other models presented in this introductory text, we have covered only the basic structure of the algorithm generating the paths of shortest distance or minimum time. There is a considerable body of literature on alternative models and on modeling extensions and refinements. They contain significant improvements over the simple approach presented here, and reduce the gap between our assumptions and the reality with which the transportation planner is confronted.

To begin with, most individuals tend to choose a path that they perceive to provide a fast connection. However, this path might not be the minimum time path, if only because users might not know which path is the fastest—frequently there are dozens of alternatives with negligible time differences. One planning strategy is therefore to distribute the projected number of trips from some origin to some destination node over a set of paths providing fast connections between them.

The path with the second fastest time, for example, can be identified as follows: Determine the minimum time path; assign to one of its links a time value of infinity; rerun the algorithm. Clearly, the new minimum time path will not be identical to the first one. Repeat this procedure m times, once for each of the m links of the original minimum time path. The fastest of the m paths generated in this way is the path with the second fastest time.

A significant part of the time it takes for trips on urban networks to reach their destination is spent waiting in front of and crossing intersections, especially if the paths include left turns. Algorithms have therefore been devised that assign time values not only to links but to network nodes as well. Others have improved on this approach by assigning turn penalties whenever a path makes a left turn when passing through a network node. All of these algorithms, however, are based on the same principles of dynamic programming discussed earlier.

6. TRANSPORTATION IMPACT

The fact that governmental agencies demand and receive environmental impact studies before approving new transportation projects does not mean that we know how to determine the impact generated by a change in a transportation system—indeed, we essentially do not. To be sure, there is a body of past experience indicating that improved transportation facilities tend to generate more traffic, more air pollution, higher land values, and changes in land use. However, our ability for quantitative prediction of these changes in space and time is almost nil.

Occasionally, history provides us with a bold example of a transportation project and some of its impacts that comes close to the setting of a laboratory experiment. Construction of the Alaska pipeline constitutes such a case. A large temporary workforce was attracted, hotel prices soared, the crime rate grew dramatically, and the U.S. dependency on foreign oil imports was reduced by some fraction, to give just a few drastic impact examples. Even in this rather conspicuous case, however, we do not know what its long-run consequences will be on the economy of Alaska or how the oil revenues will influence the quality of life in Alaska and its net migration in the future. (Worse yet, we do not even know how to measure many of the pertinent variables—for example, "quality of life.")

Impact often starts before any change in the transportation system has taken place: The mere discussion of building a new airport or extending a freeway stimulates land speculation and changes ownership and prices of land. The actual implementation of the project might require relocation of activities in the right-of-way area and generates at least a temporary infusion of additional income in the local economy. Upon completion of the project, the improved accessibility might attract new industries, which in turn leads to changes in the local commuter sheds, while other areas might experience increased unemployment resulting from the relocation of economic activities. The chain of cause-effect relationships set in motion by a change in the transportation system will interact and overlap with other changes in society and will merge with and become an undifferentiable part of the life of society and its environment.

Researchers have tried to assess the types and amounts of impact through "before and after" studies, comparing social and economic conditions before and after the implementation of a major transportation project, or they have constructed regression equations relating, say, population growth or income in a given area to a variety of independent variables, including accessibility to places of work, education, shopping, and leisure. Thus, regression equations would predict the new population or income figures resulting from a change in the accessibility, which in turn resulted from a change in the transportation system.

Predictions based on these inductive approaches are usually of low quality. For reasons essentially unknown, past events of transportation impact do not seem to permit more than a crude extrapolation to other areas and other times. The following model approaches the question of economic benefits resulting from transportation improvement by means of a deductive, deterministic strategy: It describes the interrelationships in a closed economic system by means of a set of equations and traces the consequences of a change in transportation costs through the system to identify the changes induced in production, consumption, and income. While the general theoretical framework goes back to Tinbergen (1957), the following is a simple example by Bos and Koyck (1961) that demonstrates its application.

Consider a small underdeveloped country whose national economy we will subdivide, for simplicity, into three segments that differ in location and type of production: an agricultural region, I; an industrial region, II, that refines the agricultural raw material produced in I; and another industrial region, III, producing other consumer goods. The world market is treated as a fourth economic region, IV.

A total of nine different transportation flows within the national economy and between it and the world market describe the shipments that take place among the four regions: the shipment of agricultural products from I to II, of raw materials from IV to III, of processed food from II to itself and the other three regions, and of other consumer goods from III to itself and to I and II. Shipments from IV to IV, that is, within the world market, have no bearing on the economy of the country and will not be considered. The other possible shipments are assumed to be zero: For example, there is no export from III to IV on the grounds that the products of III are not competitive on the world market, and there is, of course, no shipment from I to III, since the production of III is not based on agricultural raw materials.

The transportation costs per unit commodity from i to j are given by t_{ij}, where $i, j \in \{I, II, III, IV\}$. Of the four commodity prices p_i, those of processed food, p_2, and of imported raw material, p_4, are assumed to be fixed on the grounds that these commodities come from or are exported to the world market, which dictates their prices.

The following quantities need to be determined: incomes Y_i for the three regions of the national economy, commodity prices p_1 and p_3 of agricultural raw materials and consumer goods, the nine flow volumes v_{ij} and their monetary values V_{ij}, the production figures v_i of the three regions of our country and their respective values, V_i. A unique solution for these twenty-nine unknowns can be derived from an equal number of independent linear equations. These are:

(1) Eighteen definition equations: Each production figure v_i can be expressed by summing up the shipments from i, including the shipment from i to itself; for example, $v_3 = v_{31} + v_{32} + v_{33}$. The value V_{ij} of each shipment upon arrival is equal to its volume v_{ij} times the sum of its unit price p_i and unit transportation cost t_{ij}. The value V_i of the production at i is equal to the production volume v_i times its unit price, p_i. The income Y_i of each sector is equal to the value of its production, V_i, minus the value of raw materials received, if any. Thus, $Y_1 = V_1$, $Y_2 = V_2 - V_{12}$ and $Y_3 = V_3 - V_{43}$.

(2) Two technical equations: The volumes of the two shipments transporting raw materials to II and III are dictated by the production figures in these regions: $v_{12} = c_2 v_2$ and $v_{43} = c_3 v_3$, where c_2 and c_3 are technical input coefficients.

(3) Six demand equations: Of the nine commodity flows, two are governed by production input requirements (see the technical equations above) and one, the shipment of processed food from II to IV, is based on a fixed world market price for which this market will buy whatever quantity our country can export (perfect demand elasticity). That leaves six shipments for which explicit demand equations are needed.

Within the context of our model, the demand for commodity i at j, that is, the value V_{ij} of the shipment v_{ij}, is assumed to be a linear function of unit price and unit transportation cost of the commodity in question, of the income in j, and of the unit price and unit transportation cost of the other commodity group sold at j:

$$V_{ij} = a_{ij} Y_j + b_{ij}(p_i + t_{ij}) + c_{ij}(p_k + t_{kj}), \qquad [58]$$

where $k \neq i$ and a_{ij}, b_{ij}, c_{ij} are constant coefficients that have to be estimated empirically and are part of the input information of our model.

(4) Three supply equations: Again we assume that for the fixed price p_4, the supply of raw material from the world market is unconstrained. The supplies $v_i(i = I, II, III)$—that is, the production figures of the three

domestic regions—depend on their respective production cost/price ratios.

As an example, consider the production in the agricultural sector, I. Its unit price is p_1, and the unit production cost is assumed to depend primarily (and, in our model, exclusively) on labor wages, which in turn depend on the cost of living in that area. For the purpose of our model, unit production cost will therefore be expressed as a function of the prices $(p_2 + t_{21})$ and $(p_3 + t_{31})$ of the processed food and consumer goods in I. Thus, the following supply equation is introduced:

$$v_1 = s_0 - s_1 \frac{d_2(p_2 + t_{21}) + d_3(p_3 + t_{31})}{p_1} \qquad [59]$$

where s_0, s_1, d_2, d_3 are empirical constants. Note that the equation satisfies the usual requirement of an increase in supply with increasing unit price, and supply decrease with increasing production (that is, labor) costs. Similar equations estimate the supplies of the other two economic regions.

After linearization of nonlinear relations (for example, log transformation), the system of equations can be solved algebraically. It will yield a unique solution specifying, in numerical terms, the volumes and values of productions and commodity flows, the incomes, and the prices of the national economy.

We are now ready to utilize the model for its main purpose. Suppose that a major transportation improvement project is under consideration that will lead to reduced transportation costs $t'_{ij} \leq t_{ij} (i,j \in I, II, III, IV)$. Processing the system of linear equations with the new transportation values will yield a new solution showing the changes of production figures, of flow volumes, of prices and incomes—that is, showing, in effect, the economic growth induced by the transportation improvement.

Actual runs of this model for different sets of coefficients and functional relationships have indicated that the economic benefits (growth of national income) accruing from transportation improvement are probably substantially higher than the so-called direct benefits, defined as the transportation cost savings generated by the improvement of the transportation system.

The model can also be used as a planning instrument to achieve controlled regional growth by appropriate manipulation of selected transportation cost values. Indeed, in the form of a simulation game, the model captures two essential elements of utopian planning: the complete understanding of the dynamics of our economic environment and, based on it, the complete control of its future development.

Although transportation modeling has not yet been able to make predictions of acceptable quality about the economic consequences of transportation system changes, at least the types of impact and their measurement are known. They include changes of land values and land use, of business investments and municipal tax base, of employment and income—all measured by the usual economic scales and units.

Social impacts, on the other hand, while equally conspicuous and far-reaching, are difficult if not impossible to define and measure. Transportation and its changes may cause stress and anxiety, may provide improved access to social contacts, education, and professional careers, and may thereby change the lifestyles of neighborhoods and the social fabric of whole communities. These changes not only are the direct consequence of transportation system changes but also might result from changes in the environment at large, which in turn were caused by changes in the transportation system.

Recognizing the elusive nature of future social impacts, the transportation planners call upon the political process to deal with this uncertainty. Citizens are given the opportunity to participate in the planning process through advisory committees. Open hearings and media coverage work in favor of compromise decisions when the planning goals are controversial and future impacts uncertain; compromise decisions, in turn, tend to broaden public support and increase the base of shared responsibility once the decisions have been implemented and the impacts become apparent.

Citizen participation and public debate reflect the limits of what transportation models can contribute to the transportation planning process. If a transportation system functions relatively well, it might in part be the result of powerful transportation modeling techniques, but it might also be an expression of the shared values and the social behavior of the public that supports and utilizes the system.

REFERENCES

Beckmann, M. J. 1980. Continuous models of transportation and location revisited. *Papers of the Regional Science Association* 45: 45-53.

Bevis, H. 1956. Forecasting zonal traffic volumes. *Traffic Quarterly* 10: 207-222.

Bos, H. C., and Koyck, I. M. 1961. The appraisal of road construction projects: a practical example. *Review of Economics and Statistics* 43: 13-20.

Bruton, M. J. 1975. *Introduction to transportation planning*. London: Hutchinson.

Charnes, A., and Cooper, W. W. 1958. The stepping stone method of explaining linear programming calculations in transportation problems. *Management Science* 5: 3-8.

Chicago Area Transportation Study. 1959. *Final report I: survey findings*. Chicago: Western Engraving and Embossing Co.

———. 1960. *Final report II: data projections*. Chicago: Western Engraving and Embossing Co.

Daniels, P. W., and Warnes, A. M. 1980. *Movement in cities*. London: Methuen.

Dantzig, G. B. 1963. *Linear programming and extensions*. Princeton, NJ: Princeton University Press.

Dickey, J. W. 1975. *Metropolitan transportation planning*. Washington, DC: Scripta.

Goldman, T. A. 1958. Efficient transportation and industrial location. *Papers and Proceedings of the Regional Science Association* 4: 91-106.

Haynes, K. E., and Fotheringham, A. S. 1984. *Gravity and spatial interaction models*. Beverly Hills, CA: Sage Publications.

Hensher, D. A., and Stopher, P. R., eds. 1979. *Behavioural travel modelling*. London: Croom Helm.

Hitchcock, F. L. 1941. The distribution of a product from several sources to numerous localities. *Journal of Mathematical Physics* 20: 224-30.

King, L. J. 1984. *Central place theory*. Beverly Hills, CA: Sage Publications.

Leusmann, C. 1979. *Strukturierung eines Verkehrsnetzes* (Structuring a transportation network). Bonner Geographische Abhandlungen, Heft 61. Bonn: Ferd. Duemmlers Verlag.

Lowe, J. C., and Moryadas, S. 1975. *The geography of movement*. Boston: Houghton Mifflin.

Manheim, M. L., et al. 1975. *Transportation decision making—a guide to social and environmental considerations*. NCHRP Report 156. Washington, DC: Highway Research Board.

Martin, B. V., Memmott, F. W., and Bone, A. J. 1961. *Principles and techniques of predicting future demand for urban area transportation*. Cambridge, MA: MIT Press.

McCarthy, G. M. 1969. Multiple regression analysis of household trip generation—a critique. *Highway Research Record* 297: 31-43.

McCarthy, P. S. 1980. A general framework for the integration of a land-use model with a transportation model component. *Journal of Regional Science* 20: 51-69.

Mertz, W. L., and Hamner, L. B. 1957. A study of factors relating to urban travel. *Public Roads* 9: 13-20.

O'Kelly, M. E. 1983. Multipurpose shopping trips and the size of retail facilities. *Annals of the Association of American Geographers* 73: 231-39.

O'Sullivan, P., Holtzclaw, G. D., and Barber, G. M. 1979. *Transportation network planning*. London: Croom Helm.

Potts, R. B., and Oliver, R. M. 1972. *Flows in transportation networks*. New York: Academic Press.

Quandt, R. E., and Baumol, W. J. 1966. The demand for abstract transport modes: theory and measurement. *Journal of Regional Science* 6: 13-26.

Ralston, B. A., and Barber, G. M. 1984. Taxation and optimal road penetration. *Geographical Analysis* 16: 313-30.

Scott, A. J. 1971. *Combinatorial programming, spatial analysis and planning*. London: Methuen.

Smith, T. R., and Slater, P. B. 1981. A family of spatial interaction models incorporating information flows and choice set constraints applied to U.S. interstate labor flows. *International Regional Science Review* 6: 15-31.

Stopher, P. R., and Meyburg, A. H. 1975. *Urban transportation modeling and planning*. Lexington, MA: D. C. Heath.

———. 1979. *Survey sampling and multivariate analysis for social scientists and engineers*. Lexington, MA: D. C. Heath.

Stouffer, S. A. 1960. Intervening opportunities and competing migrants. *Journal of Regional Science* 2: 1-26.

Tinbergen, J. 1957. The appraisal of road construction: two calculation schemes. *Review of Economics and Statistics* 39: 241-49.

U.S. Department of Transport. 1977. *National transportation: trends and choices*. Washington, DC: U.S. Government Printing Office.

Wendt, P. F., ed. 1976. *Forecasting transportation impacts upon land use*. Boston: Martinus Nijhoff.

Werner, C. 1968. The law of refraction in transportation geography: its multivariate extension. *Canadian Geographer* 12: 28-40.

Wheeler, J. O. 1974. *The urban circulation noose*. North Scituate, MA: Duxbury Press.

Wilson, A. G., et al. 1981. *Optimization in location and transport analysis*. New York: John Wiley.

Wootton, A. J., and Pick, G. W. 1967. A model for trips generated by households. *Journal of Transport Economics and Policy* 1: 137-153.

Yeates, M. 1963. Hinterland delimination: a distance minimizing approach. *Professional Geographer* 15: 7-10.

Yule, G. U., and Kendall, M. G. 1965. *An introduction to the theory of statistics*. New York: Hafner.

ABOUT THE AUTHOR

CHRISTIAN WERNER, Professor of Social Sciences at the University of California at Irvine, in traditional European fashion did his undergraduate work by attending six universities. In 1960, he received an M.A. in mathematics and an M.A. in geography from the University of Berlin. He studied geography for two years at Northwestern University and in 1965 returned to Berlin, where he presented his dissertation and accepted an appointment as Assistant Professor. The following year, he joined the faculty at Northwestern University. In 1969, he became Professor of Social Sciences at the University of California at Irvine. In 1972, he held the Wallace J. Eckert Science Chair at the I.B.M. Research Center in New York. He returned to Irvine to become Dean of Social Sciences. In 1979, he was Visiting Professor at University of Lagos, Nigeria. A sample of his many articles can be found in a diverse set of journals, including *Ekistics, Annals of Regional Science, Geographical Analysis, Canadian Geographer, Nigerian Geographical Journal,* and *Water Resources Research.* He is a member of the Panel on Methodological Research Frontiers and the Social Sciences for the National Science Foundation, and Chairman of the President's Advisory Council for the Institute of Transportation Studies for the University of California Statewide System.